Praise for *The Art of Positive Politics*

"A nearly flawless treatment of some of the most vexing, but critically important, aspects of the project manager's job—managing power and political behaviors in positive ways designed to move their projects to successful completion. I cannot think of an author with greater credibility to write this book than Vijay Verma."

—Jeffrey K. Pinto, PhD
Black Chair in the Management of Technology
Black School of Business, Penn State University

"Irrespective of whether you are a novice or an expert in project management, you'll find this book highly valuable. Vijay offers keen insights and practical tips to help you navigate the labyrinth of politics."

—Dr. Prasad S. Kodukula, PMP, PgMP
Founder & CEO, Kodukula & Associates, Inc.
Adjunct Faculty, Project Management, University of Chicago & Illinois Institute of Technology

"Everyone is in a political environment—both internally and externally; we cannot escape it. Vijay Verma explains how we can best use politics to our advantage in a must-read book. I recommend it highly."

—Dr. Ginger Levin, PMP, PgMP, OPM3

"In this latest book by well-established author Vijay Verma, Vijay aims directly at the most beguiling aspect of project management—handling the challenges of corporate politics in promoting projects, especially the typically negative reactionary responses to a new project launch. Why do negative politics transcend positive politics? Because if you are comfortable with a project's planned impacts and outcomes, no need to say anything. After all, it is not your job to promote a project unless you are specifically assigned to do so.

"But the project manager's job is to turn that situation around. In The Art of Positive Politics, *Vijay explains how to analyze the political landscape, understand stakeholders' political positions from their perspective, and thereby convert opposition into support that results in even more successful results.*

"For those project managers appropriately inclined, managing the political environment can be an exciting and satisfying component of the project manager's job, especially when it enhances the ability to get things done. The best part of all? Vijay covers the broad landscape from the project level up to the top organizational level. Highly recommended."

—R. Max Wideman, FCSCE, FEIC, FICE, FPMI, FCMI
(Retired owner of www.maxwideman.com)

....

"In a world where "politics" is often a dirty word, Vijay finds ways to help others exercise their political muscle while driving home the positive. He moves the political ball from "I" to "We" with aplomb."

—Carl Pritchard, PMP, PMI-RMP
Principal, Pritchard Management Associates

....

"Delivering projects on budget and on time requires the cooperation of many people — and that is an inherently political process. Mastering politics — positive politics — is both an art and a science, and in this engaging and insightful book, Vijay Verma describes the lessons he learned applying the art of positive politics to project management at TRIUMF, Canada's particle accelerator center. Vijay's astute observations apply to projects both large and small, and are required reading for successful project managers – and organizational leaders alike."

—Dr. Jonathan Bagger
Director, TRIUMF, Canada's Particle Accelerator Center

....

"Vijay has been delivering workshops and sessions at the ProjectWorld and ProjectSummit events for over twelve years and consistently delivers exceptional educational value to the attendees. He is fondly referred to as the 'guru' of Project Management: a fitting title indeed! I would recommend this book to leaders and management who are passionate inspiring and fostering positive communication and collaboration within their organizations."

—Amy Ruddell
Senior Director, Project World/BA World and Project Portfolio, Macgregor Communications
Editor-in-Chief, ProjectTimes and BATimes

....

"Having known and worked with Vijay over many years, I have found his books, workshops, and support services exceptional in value and actionable ideas they provide to those who work in project management."

—Dave Po-Chedley

....

..."Vijay has again mastered a practical and on-the-point approach to project solutions—a must-have."

—Jens Dilling, PhD
Professor at University of British Columbia and Associate Director, Physical Sciences Division, TRIUMF, Canada's National Laboratory for Particle and Nuclear Physics.

....

"Verma continues to lead our understanding of the uncertain political landscapes of projects. He turns an often negative topic into a positive opportunity to increase project success."

—Dr. Dale Christenson, DPM, PMP, CMC
President, Project Management Centre of Excellence Inc.

....

"Like it or not, politics are an inevitable part of project and organizational life. Ignore or mismanage it at your own peril. Vijay has taken a relatively complex topic and presented it in a comfortable and easily digestible manner for all of us. A key message is the power we can wield by demonstrating positive politics. This book should be in everyone's toolbox as early in their career as possible."

—Neal Whitten, PMP
Principal, The Neal Whitten Group

....

"Delivering projects on budget and on time requires the cooperation of many people — and that is an inherently political process. Mastering politics — positive politics — is both an art and a science, and in this engaging and insightful book, Vijay Verma describes the lessons he learned applying the art of positive politics to project management at TRIUMF, Canada's particle accelerator center. Vijay's astute observations apply to projects both large and small, and are required reading for successful project managers – and organizational leaders alike."

—Dr. Jonathan Bagger
Director, TRIUMF, Canada's Particle Accelerator Center

....

"Vijay has been delivering workshops and sessions at the ProjectWorld and ProjectSummit events for over twelve years and consistently delivers exceptional educational value to the attendees. He is fondly referred to as the 'guru' of Project Management: a fitting title indeed! I would recommend this book to leaders and management who are passionate inspiring and fostering positive communication and collaboration within their organizations."

—Amy Ruddell
Senior Director, Project World/BA World and Project Portfolio, Macgregor Communications
Editor-in-Chief, *ProjectTimes* and *BATimes*

....

The Art of Positive Politics

A Key to Delivering Successful Projects

by Vijay K. Verma
PMI Fellow, PMP, MBA, PEng

First Edition

Multi-Media Publications Inc.
Oshawa, Ontario

The Art of Positive Politics: A Key to Delivering Successful Projects

by Vijay K. Verma

Managing Editor:	Kevin Aguanno
Copy Editor:	Susan Andres
Typesetting:	Peggy LeTrent & Charles Sin
Cover Design:	Arrow Designs
eBook Conversion:	Sebastian Aguanno

Published by:
Multi-Media Publications Inc.
Box 58043, Rosslynn RPO, Oshawa, ON, Canada, L1J 8L6.
http://www.mmpubs.com/

Paperback   ISBN-13: 978-1-55489-177-1
eBook Formats   ISBN-13: 978-1-55489-178-8

Published in Canada. Printed simultaneously in Canada, the United States of America, Australia and the United Kingdom.

CIP data available from the publisher.

Dedication

To Shiksha, love of my life, my partner, and my best friend, for her continuous and unconditional support;

To my late parents and uncle (Taya Ji) for their inspiration;

To our grandchildren, Nikhil, Neel, Rohnik, Jaya, Reyva, Rahi, and Veeyan for their love and smiles and for teaching me the importance of listening to better understand and empower one another toward achieving extraordinary results together.

The Art of Positive Politics

Table of Contents

Table of Illustrations 13

Foreword 19

Preface 25

Acknowledgments 29

Introduction 33

Part I: Politics & Project Management .. 41

1: Basic Concepts of Politics 45

2: Politics in Project Management 67

Part I Summary 91

Part II: Dynamics of Politics 95

3: Two Types of Politics *97*

4: Why Organizational Politics? *137*

5: Politics in Managing Stakeholders *165*

Part II Summary *203*

Part III: Understanding the Political Landscape & Stakeholder Behaviors ... 207

6: Analyzing the Political Landscape *211*

7: Understanding Political Behaviors *243*

8: Managing Political Behaviors *275*

Part III Summary *305*

Part IV: The Art of Managing Politics .. 311

9: Three Truths of Life to Manage Stakeholders . *315*

10: Managing Politics at the Management Level *331*

11: Managing Politics at the Project Level *365*

Part IV Summary *415*

Appendix: Case Studies & Exercises 419

End Notes 433

References 451

Index 463

About the Author 487

The Art of Positive Politics

Table of Illustrations

Figures

Figure 1.1. Intensity of politics in a project management environment.

Figure 1.2. Power and politics in project management.

Figure 1.3. From strategies to results with power, politics, and leadership.

Figure 2.1. Level of politics according to project life span.

Figure 3.1. Commitment and motivation model.

Figure 4.1. Four main issues leading to politics.

Figure 5.1. Matrix showing levels of power and predictability.

Figure 5.2. Matrix showing levels of power and commitment to change.

Figure 5.3. Matrix showing levels of power and respect.

Figure 5.4. Matrix showing levels of power and ethics.

Figure 5.5. Matrix showing levels of power and integrity.

Figure 6.1. Swimming in a political pond.

Figure 6.2. Politically Sensibles' priority sequence for using politics.

Figure 6.3. Conflict management styles based on time and standards of fairness.

Figure 9.1. Three truths of life.

Figure 9.2. Gaining support through genuine involvement.

Figure 11.1. Relationship between project complexity level and positional power of a champion.

Figure 11.2. Five categories of stakeholders (based on trust and agreement). Adapted from Peter Block, *The Empowered Manager: Positive Political Skills at Work* (San Francisco: John Wiley and Sons, Inc., 1987), 139; used with permission.

Tables

Table 1.1. Eight Sources of Power (Types and Bases)

Table 2.1. Emotional Views of a Troubled Project

Table 2.2. Survival Suggestions for Political Phases

Table 2.3. Proactive Views of a Troubled Project's Phases

Table 2.4. Advantages of Proactivity for Political Phases

Table 3.1. Attributes of Negative Politics and Their Purposes

Table 3.2. Ten Commandments of Positive Politics and Their Purposes

Table 3.3. PRIDE (in a Nutshell)

Table 4.1. Major Issues in Developing Good Champions, with Recommendations

Table 5.1. Matrices Showing Levels of Power and Category-1 Attributes of Stakeholders

Table 6.1. Political Viewpoints and Management Techniques of Flounders (Naives). Adapted from Jeffrey K. Pinto, *Successful Information System Implementation: The Human Side, The Perspective Series* (Upper Darby, PA: PMI®, 1994), 118–122; and Jeffrey K. Pinto, *Power and Politics in Project Management* (Newton Square, PA: PMI, 1998), 75–77.

Table 6.2. Political Viewpoints and Management Techniques of Sharks. Adapted from Jeffrey K. Pinto, *Successful Information System Implementation: The Human Side,*

The Perspective Series (Upper Darby. PA: PMI®, 1994), 118–122; and Jeffrey K. Pinto, *Power and Politics in Project Management* (Newton Square, PA: PMI, 1998), 75–77.

Table 6.3. Political Viewpoints and Management Techniques of Dolphins (Politically Sensibles). Adapted from Jeffrey K. Pinto, *Successful Information System Implementation: The Human Side, The Perspective Series* (Upper Darby, PA: PMI®, 1994), 118–122; and Jeffrey K. Pinto, *Power and Politics in Project Management* (Newton Square, PA: PMI, 1998), 75–77.

Table 6.4. Characteristics of People Engaged in Negative Politics

Table 7.1. Political Behaviors and Impact at the Interpersonal Level

Table 7.2. Political Behaviors and Impact Related to a Team Environment

Table 7.3. Political Behaviors and Impact Related to the Organization

Table 8.1. Guidelines for Managing Political Behaviors at the Interpersonal and Team Levels

Table 8.2. Guidelines for Managing Political Behaviors at the Organization Level

Table 10.1. Managing Politics at the Management Level (Organizational Issues)

Table of Illustrations

Table 10.2. Managing Politics at the Management Level (Leadership Issues)

Table 10.3. Managing Politics at the Management Level (Project Management Issues)

Table 11.1. Managing Politics at the Project Level (Political Issues and Managing Upward)

Table 11.2. Managing Politics at the Project Level (Project Management and Team Leadership Issues)

Table 11.3. Managing Politics at the Project Level (Stakeholder Management Issues)

The Art of Positive Politics

Foreword

I've lived in the Washington, DC, area for several decades. Our number one industry here is politics. Although I have never been politically active, a consequence of living here so long is that I know how politics work, what's good about them, and what's bad. Without trying, I have become a bit of an expert on politics through osmosis. Sadly, at the time of this writing, the bad of politics dominates DC. Recent public opinion surveys carried out by the Gallup organization rank members of Congress last in trustworthiness in a list of twenty-one professions, below car salespeople and lobbyists.

The good news is that at some point, the pendulum will swing back. I have witnessed the periodic swings of politics from good to bad and back to good again because I have lived in Washington so long, and my experience tells me that the political dysfunctionality we currently experience will be replaced by productive political collaboration again—hopefully sooner, rather than later. However, even when good politics dominate, we need to know the undercurrents of the bad, and when the bad dominates, we take solace in finding traces of the good.

In his book, *The Art of Positive Politics*, Vijay Verma shines light on the two faces of politics. The first face—the ugly, sour one—captures the negative mien of politics—bullying, dissimulation, backstabbing, and circumventing what is right. When people think about what politics are about, most carry this image in their heads. When a colleague tells you, "Our director is a real politician," the sentiment is not offered as a compliment.

The second face of politics—earnest, capable, trustworthy—reflects politics as positive phenomena. To the extent politics strive to get teams, organizations, and communities to support initiatives that serve the greater good, it plays an important positive role. Good politics achieve beneficial goals by building consensus. A key component of reaching consensus is influencing outcomes by addressing the players' concerns and needs. When carried out positively, politics bring together individuals who have different backgrounds and capabilities, and who might otherwise pursue their agendas dysfunctionally.

In this book, Mr. Verma explores the dual nature of politics with a view to developing insights that apply to managing projects. Taking a page from Robert Block's classic 1983 work, *The Politics of*

Projects,[1] he argues that project environments are inherently political, because even simple projects are composed of multiple players who hold different views and pursue often-misaligned goals. If managers are going to handle the centripetal forces that can drive project work this way and that, ultimately paralyzing progress, they need to take stock of the range of players who can have an impact on the project effort, as well as people the project will affect. They must understand these players' interests so they can be addressed.

This perspective is explored more deeply in Lee Bolman and Terrence Deal's classic book on how organizations function, titled *Reframing Organizations* (2013).[2] They argue that to understand how organizations work, you need to look at them in accordance with four frames: structural, human relations, symbolic... and political. Yes, the political frame is given equal billing with traditional approaches to understanding the functioning of organizations. The inclusion of the political dimension in examining organizational operations was novel when it first appeared, because the traditional treatment of organizational behavior principally focused on structural issues (organizational design) and human relations issues (people as the key players in organizations). The political dimension was overlooked.

Bolman and Deal believe that understanding behavior's political dimension in organizations is important. Like Mr. Verma, they point out that organizational politics are not inherently good or bad. However, they are inevitable, which offers a compelling reason to understand them. Politics arise because in organizations, different people hold different views on a host of matters, and they form interest groups to promote these views. They form coalitions to add numbers to their perspective. They understand that to resolve disputes, they must engage in negotiations and develop solutions that entail compromise. In short, they engage in political action.

In articulating his views on the political aspect of project management, Mr. Verma draws on his extensive experience in studying the people dimension in managing projects and writing and lecturing on this topic extensively. He is an important thought leader on the subject. His first three books on people in project management— *Organizing Projects for Success* (1995),[3] *Human Resource Skills for the Project Manager* (1996),[4] and *Managing the Project Team* (1998)[5]—established his *bona fides* as an expert. In recognition of his contributions to understanding the human side of project management, Vijay Verma was made a Fellow of the Project Management Institute in 2009—a highly distinguished award. In a professional society of nearly a half-million members, only forty-five men and women hold the title of PMI Fellow.

Mr. Verma is not resting on his laurels now that he has completed *The Art of Positive Politics: A Key to Delivering Successful Projects*. He is busily at work on a follow-up volume that deals with the popular topic of influencing without authority. One trait of being a project manager is that because you are typically dealing with borrowed resources (matrix management), you have little or no authority over anyone in the project environment. You certainly do not have authority over your boss or the

project steering committee or external stakeholders. What is a bit unnerving is that, generally, you do not even have authority over the team you manage, because these people are often borrowed resources.

So, how do you get anything done, when you find yourself in a position where you cannot command outcomes from people who are working on or otherwise supporting the project? The simple answer: You need to be good at *influencing* these players. Inasmuch as politics are often defined as "the art of influence," you will find that you are playing the role of politician on your project.

This is the perspective Vijay Verma will address in the follow-up book. I hope that by the time the book is published, the pendulum will have swung from today's political dysfunctionality in Washington to productive civility in politics. Even then, we recognize that pendulums being pendulums, at some point, it will swing back in the direction of the bad in politics. And so it goes.

— J. Davidson Frame, PhD, PMP, PMI Fellow
Academic Dean, University of Management & Technology
Arlington, VA (July 4, 2017)

The Art of Positive Politics

Preface

When "I" is replaced by "We," even
illness becomes wellness! Similarly,
negative politics focus on "I," and
positive politics focus on "We."

          — Anonymous

Politics are a popular and emotionally hot topic. Over the many years of my practical experience, I noticed that politics derail many projects, despite good planning and scheduling. I talked with many project managers and learned that most dislike politics and consider them negative and wasted time. Few understand the importance and dynamics of politics and recognize that politics are inevitable and necessary to get things done through project stakeholders.

Generally, politics have a negative connotation, but I present a practical view of politics and present ideas of leading with the art of positive politics to deliver successful results. Politics cannot be eliminated; therefore, project managers must learn to manage politics. They place high priority on developing skills for navigating organizational politics to manage their projects effectively only after many scars. Successful leaders recognize that power, influence, and politics are interrelated, and all three are important to achieve extraordinary results in any organization.

Many program and project managers encounter the following topics:

- Ignoring the external and internal politics surrounding your project can be hazardous to success.

- How can you create and use positive politics to minimize the negative impact of politics?

- How do you analyze the political landscape to understand and manage stakeholders' political positions and behaviors?

- Good leadership and political savviness are critical to achieve better results.

I like to share my knowledge and experience with others because knowledge is power, and the more you share it, the more you get. Books are excellent media to achieve this because they allow me to reach many people I might be unable to meet in person.

I first thought of this book while writing a chapter, "Leadership, Power, Influence, and Politics in Project Management," in *Human Resource Skills for the Project Manager* (1996).[1] That book was the second in my series of three books on the Human Aspects of Project Management published by PMI (Project Management Institute). I tabled the idea at the time, but over subsequent years as I taught seminars, I learned from my students that the topic is much broader and deeper, and research showed me that the existing project management literature, including *A Guide to the Project Management Body of Knowledge* (*PMBOK® Guide*) (2013),[2] does not cover it sufficiently.

Comments from many people about the lack of political skills motivated me to write a book to help managers in all industries understand the dynamics of politics and develop practical skills to manage politics at various levels. Therefore, this book presents guidelines to help portfolio managers, program managers, and project managers deliver successful results to meet organizational strategies and goals.

This book should help senior management minimize the negative impact of politics by creating positive politics characterized by the attributes, keywords, and ideas described. Portfolio, program, and project managers will learn to analyze their political landscape to understand three main political positions (Naives, Sharks, and Politically Sensibles) and stakeholders' various political behaviors.

Academics and educators in business schools and project management programs can use this book to develop a course to teach practical ideas about power, influence, and politics. The concepts, ideas, and guidelines presented apply to initiatives and projects in any industry—government, information technology, construction, oil and gas, utilities, transportation, aerospace, mining, pharmaceuticals, research and development, financial, banking, insurance, hospitality, arts, and service industries.

Acknowledgments

Writing this book was a new challenge for me. As my first big project after retirement, I was distracted by wanting to spend more time with my grandchildren; so, it required self-motivation, perseverance, and discipline. Most of us find it difficult to adopt these characteristics, and I am no exception. However, I am fortunate to have my family and many others in my professional life who encouraged me to stay focused and supported me throughout this project.

I am grateful to my special friend and mentor, R. Max Wideman, for his moral support and encouragement in writing this book that revolves around the unique idea of positive politics. I also thank participants from my classes and seminars on the human aspects of project management for their many ideas and discussions that helped frame my thoughts for this book. These discussions inspired me to highlight and recognize the positive side of politics to help managers in all industries analyze the political landscape and manage politics effectively at various levels.

I am indebted to Kevin Aguanno of Procept Associates Ltd. He acknowledged that politics have a negative connotation, and my idea of presenting a practical view of politics was unique and exciting. He suggested, based on his many years of practical experience in project management and publishing, that there are many new interesting concepts and excellent ideas I should compile in my next projects and series of books.

I extend my sincere thanks to Raso Samarasekera who transcribed my draft manuscript and went through the several revisions cheerfully. I also extend this thanks to Des Ramsay and Shirley Ramsay for their help in editing, as well as Adrian Watt, Mark Keyzer, and Larry Lee for their help in preparing the figures and tables.

The person who pushed me to the finish line for this book was Susan Andres, an amazing editor at Multi-Media Publications, Inc. She edited my manuscript professionally and helped me convey my messages and ideas much more clearly and precisely. I appreciate the many good discussions we had that kept this book practical and perhaps easier to understand. I cannot express in words Susan's extraordinary help throughout the journey of writing this book, but I am thankful that she did it with the highest standards of quality and cheer.

My wife, Shiksha, is an unintentional contributor to this book at a level at least equal to me. She contributed deep insights, thought-provoking ideas, and many lengthy discussions that helped formulate much of the content. If I acknowledge her individual contributions, her name would literally be on every page.

Finally, I am grateful to my son, Naveen, my daughters, Serena and Angelee, and their spouses, Anita Misri, Jaimini Thakore, and Krupal Patel, who encouraged me, offered great suggestions about better understanding people, and gave me the freedom to complete this book.

Introduction

This book focuses on the concepts of politics and on guidelines to manage politics at all levels in delivering successful projects. It covers two types of politics (negative and positive, with associated attributes for each), analyzing the political landscape to identify three political positions of stakeholders (Naives, Sharks, and Politically Sensibles), understanding and managing political behaviors, and managing politics at the upper management level as well as at the project level.

Most managers and leaders have experienced the following realities in the corporate world:

- People are *hired* for good technical skills.

- People are *promoted* for good interpersonal skills.

33

- People are *fired* because they lack good leadership skills.

Politics are an important part of life in project management environments because projects are done by people with different viewpoints, expectations, interests, and personalities. People make things happen and prevent things from happening. Therefore, wherever there is a diverse mix of people, there are politics.

Politics exist in every organization at all levels, and the intensity of politics varies by the project's complexity. The more complex the project is, the higher the intensity of politics is. In this context, complexity depends on how many functional departments are involved in doing the project because each department has its preferred ways of doing things. The intensity of politics is also generally high at the top, and it varies as management of organizational strategies breaks down into management of portfolios, programs, and projects.

After discussions with many project management practitioners, I discovered most project managers encounter the following questions and issues:

- What are the importance and dynamics of politics, and can politics be positive?

- How do you analyze the political landscape and understand political positions and stakeholder behaviors?

- How can you convert your opponents and adversaries into allies to build support for your vision?

- How do you survive and progress when swimming with Sharks without being eaten alive in an organizational political pond?

- How can you manage politics effectively at the upper management level and at the project level?

- How can you refine influencing and leadership skills to get extraordinary results?

Power, influence, and politics are important to produce high performance. Power and politics are inseparable in that power is an ability to influence others to do certain things, and it is necessary to navigate organizational politics successfully. Politics are associated with how people use their power to manage politics effectively and get things done to deliver successful results. In a way, politics are about seeking power, acquiring power, and maintaining power by using it appropriately.

Project managers generally dislike organizational politics. There is not even a universal definition of politics because their dynamics are complex. Politics can be both negative and positive. Negative politics emphasize **I**, whereas positive politics emphasize **we**. Positive politics focus on how we deal with one another effectively to help the organization. Positive politics are least understood, and many managers at various organizational levels view them with suspicion and uncertainty. I developed a set of ten main ideas behind positive politics, which may be considered Ten Commandments to powerfully minimize the negative impact of politics:

1. Establishing and communicating organizational strategies and goals to create unity of purpose

2. Building honest and long-term alliances

3. Encouraging collaboration rather than competition

4. Creating synergy to create high-performance teams

5. Distributing rewards without favoritism

6. Inspiring quality of results

7. Providing support at the professional and personal levels

8. Developing and implementing a strong mentorship program

9. Sharing information to empower all stakeholders

10. Embracing cultural diversity

Because politics cannot be eliminated, I suggest using a combative approach by creating positive politics to minimize the impact of negative politics. The main aim of positive politics is to redirect the energy of people involved in negative politics toward positive aspects of politics to create a culture of more collaboration, synergy, and commitment. Program and project managers should view politics positively and use them to gain support and cooperation from project stakeholders to meet organizational strategies and goals.

Today's business environment is characterized by global competition, rapidly changing technology, and limited human resources with the proper skill mix. Project management is critical to business organizations' success to transform their strategies and goals into an appropriate set of portfolios, programs, and projects. Management by Projects (MBP) is becoming a successful way to manage an organization. Effective project management integration requires combined project management skills, industry-specific skills, and people skills. The people skills include communication, motivation, negotiation, conflict management, team building, and leadership. Even though the concepts, dynamics, and impact of power and politics are interesting and critical to deliver successful projects, they are not discussed in depth in project management literature.

Learning Objectives

This book is divided into four parts. Part I presents basic concepts and builds a strong foundation with practical examples to achieve people's potential, emphasizing that politics are about power. The role of politics in project management is introduced with an explanation of how politics intensify over the project life span (PLS) and a look at the dynamics of a troubled project from emotional and proactive viewpoints.

Part II explores the dynamics of politics and the two types of politics (negative and positive), describing the attributes of negative politics and the Ten Commandments of positive politics. It also covers the reasons for organizational politics and the politics involved in managing stakeholders. Part

III delves into analyzing the political landscape and understanding and managing stakeholder behaviors. Finally, Part IV looks at the three truths of life in managing stakeholders and strategies to manage politics at the management and project levels.

The topics in all chapters have been organized to help readers find information relevant to their interest. The concepts and ideas are clearly explained and illustrated by figures, and the main points are highlighted in lists.

After reading this book and relevant reference materials, readers will better understand the following:

- Dynamics and complexity of politics and relationship between power and politics

- Six phases of a troubled project with issues and solutions related to power and politics (the emotional side and proactive side)

- Importance of people (project champions), people (project managers), people (team members) to deliver successful results

- Two types of politics (negative and positive) with attributes, keywords, and ideas

- Various reasons for politics associated with project management issues, people behavioral issues, global market issues, and politics in managing stakeholders

- Analysis of the political landscape, identifying three political positions of stakeholders (Naives, Sharks, and Politically Sensibles) with their viewpoints about politics, intentions to deal with politics, and strategies to manage their people and projects

- Understanding and managing political behaviors of various stakeholders at the interpersonal level, team level, and organizational level

- Three truths of life as a key to manage politics

- Strategies to manage politics at the senior management level (to resolve organizational issues, leadership issues, and project management issues)

- Strategies to manage politics at the project level (to resolve political issues and challenges of managing upward, team leadership and project management issues, and stakeholder management issues)

Part I. Politics and Project Management (Setting the Stage)

Chapter 1: Basic Concepts of Politics

1.1. Definition and Dynamics of Politics

1.2. Why Politics Are Complex

1.3. Two Components of Politics

1.4. Politics Are About Power

Chapter 1 Summary

Chapter 2: Politics in Project Management

2.1. Project Life Span (PLS) and Politics

2.2. Dynamics of a Troubled Project (the Emotional Side)

2.3. Dynamics of a Troubled Project (the Proactive Side)

Chapter 2 Summary

Part I Summary

Politics and Project Management (Setting the Stage)

Politics are the science of who gets what, when and why.

—Sidney Hillman

In the opening quote, Sidney Hillman says politics are a *science*, but I believe they are the *art* of getting things done—why some are better than others are in understanding the dynamics of politics and managing them effectively to deliver successful results.

Management by Projects (MBP) is becoming one successful way to manage an organization. Therefore, project management, a technique of getting things done and delivering specified objectives, is essential to a business organization's success. It involves using combined technical skills based on *A Guide to the Project Management Book of Knowledge* (*PMBOK® Guide*) (2013)[1] and soft skills.

41

Power and politics are least discussed among soft skills but most important to deliver successful projects. Power is the ability to influence stakeholders to complete project activities assigned them, and politics are associated with how people use their power to get things done and manage their projects. Project managers must recognize that power and politics exist at all organizational levels, and their dynamics are complex enough to affect project success significantly. Therefore, project managers must consider power and politics necessary in project management and develop skills to manage politics effectively.

Many projects are derailed because of politics, despite good planning and project management processes. Politics are inevitable in project management environments, regardless of whether project managers like them. Politics occur because stakeholders have diverse backgrounds, expectations, and ideas of how things should be done to manage projects. These politics might be related to how the projects are planned, organized, and managed. A common issue relates to how and why few people get the best resources. Project managers get upset because they do not get enough organizational support for their projects, whereas those with even lower priority get the resources they want.

Although organizational politics play an important role in determining a project's success or failure, we understand little about power and politics in project management. Much anecdotal evidence and many case studies show the impact of politics in project environments. Therefore, project managers must recognize they will encounter politics and, therefore, must learn to navigate through them to deliver successful projects.

Power and politics are interrelated, which sometimes makes understanding and managing politics complex and challenging. Part I deals with the basic concepts of politics that cover their definition and dynamics. It describes how using power and influence effectively helps manage politics. Although politics have a negative connotation, they can be both negative and positive.

The Art of Positive Politics

Basic Concepts of Politics

*Politics are the conduct of public affairs
for private advance.*
—Ambrose Bierce

Politics have existed everywhere in every organization for ages. Project managers must deal with diverse stakeholders with different backgrounds and expectations. In a project environment, they must interact with many departments with their way of doing things. People feel uncomfortable when someone from outside tries to impose ideas and processes on them to do things differently. They resent processes and methodologies they are not used to using in their departments.

People do projects. People make things happen and prevent things from happening in project environments. People are interested in increasing their power so they can get what they want when they want it, even if others oppose it, including getting resources and support for a vision or a process, organizational change, or a new method. Moreover, people are complex, unpredictable, and driven by their intrinsic motives and thoughts, making politics complex and often not transparent.

Power and politics are important topics in project management, but they are not discussed in depth in the *PMBOK Guide*. This book discusses these topics in detail to develop better understanding of the dynamics of power and politics and their effective use in managing projects successfully. Although power only provides the ability to influence project stakeholders, appropriate use of power leads to long-term and positive influencing. Just having power is insufficient; knowing how to use power effectively is the key to getting things done. It helps navigate organizational politics, leading to successful project management.

This chapter deals with the definition of politics and people's different views of politics. Politics relate to power and influence. The dynamics of power are complex, and influencing is individualized. I cover how and why power is complex and how power is necessary to manage politics effectively to deliver successful project outcomes. You will learn the three necessary ingredients of power and two components of politics.

1.1. Definition and Dynamics of Politics

Wherever there are people, there will be politics. Politics can make or break project success.

—Vijay Verma

Politics are a popular and emotionally hot topic, and project managers disagree about the notion of organizational politics. Politics have no universal definition, and their dynamics are complex. Politics imply a diplomatic way to manage people and projects in an organization. Some people think politics are a science, but they are more of an art—the complex art of dealing with people to get something you want when you want it. Sometimes, they include following specific directions and making changes to deliver project outcomes. They involve influencing stakeholders to work toward your goals and objectives.

Politics consist of unwritten and unspoken ways to get things done. They make people anxious and sometimes upset because some people get what they want without going through proper steps and procedures others must follow. However, this is a reality of life in the business world and, especially, in project environments where resources are limited. I determined from my many presentations in seminars and workshops on politics that people have the following ideas (Verma 2008–2015):[1]

- Politics are hidden and personal agendas.

- Politics are about favoritism.

- Politics are about power struggles, turf wars, and empire building.

- Politics are bad and unpleasant.

- Politics are negative.

- Politics are the way to push your ideas and ways to do things.

- Politics are unspoken rules and ways to get things done.

- Politics are about satisfying personal ego.

- Politics represent invisible undercurrents.

- Politics are a way to make people do things your way.

- Politics are all about relationships—whom you know and how well.

- Politics involve negotiations behind the back doors.

- Politics are about the "good old boys network."

- Politics are about getting what you want by using your network.

Considering these viewpoints, many people dislike politics because they have a negative connotation. A project environment has a finite time and budget to deliver specified outcomes for a product or service. Project managers are under pressure to meet milestones. They often encounter behaviors from project personnel that they don't understand and endorse. They think those behaviors and disagreements are not clearly expressed and based on self-interests and ulterior motives rather than the organization's interests. Such actions might be called political. However, not all politics are negative. Positive aspects of politics create a culture of more collaboration and teamwork.

1.2. Why Politics Are Complex

*Politics are complex because
people make them so.*

—Vijay Verma

Most projects are organized in a matrix, where people from different departments, with diverse skills and backgrounds, work as a project team. Project managers often have great responsibility, but no direct formal authority over their team members and other project stakeholders. Therefore, they must learn to use *influence* rather than *command and control*. Power enables them to influence project stakeholders to do project activities on time and within budget to meet milestones and deliver a successful project.

Politics are complex and difficult to understand and manage because people are complex and difficult to understand. Power is an important part of politics that can be derived from various sources. Power is also based on perceptions seldom specified clearly. Power's intricate nature makes the dynamics of politics even more complex.

Understanding the dynamics of power and politics is important so organizational leaders can rechannel people's energy in a positive direction, develop a culture of more collaboration and synergy, and develop PPM process and methodology to meet organizational strategies and goals. Figure 1.1 shows how the intensity of politics relates to management of projects, programs, and portfolios in most organizations. The intensity of politics consists of the following four variables:

1. Degree of interaction among stakeholders

2. Frequency of interaction among stakeholders

3. Competing requirements and agenda of stakeholders

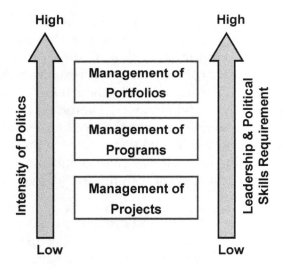

4. The stakeholders' desire to reach an agreement

Figure 1.1. Intensity of politics in a project management environment.

Whenever and wherever people are involved, there are politics. Figure 1.1 shows that project managers face the challenges of managing with a level, or intensity, of politics, that increases as they manage projects, programs, and then portfolios and as more people with diverse interests from different departments are involved and expected to work in teams.

The need to align with business strategies and optimize resource constraints (quantity and skill mix) becomes more important. Portfolio managers must develop their negotiating and people skills to influence stakeholders from different departments. They must increase their informal power to navigate politics positively and enhance their leadership skills to achieve synergy among stakeholders across the organization.

1.3. Two Components of Politics

In politics, absurdity is not a handicap.
—Napoleon Bonaparte

Politics are inevitable when managing projects. Project managers must understand the dynamics of politics to meet challenges when dealing with stakeholders. Politics have two components:

1. Tangible and obvious – Published and written policies and procedures project personnel are expected to follow in their daily lives.

2. Intangible and unobvious – Unspoken rules and unknown ways to get things done, including *network power*, that is, whom you know,

how to use those connections effectively, and what knowledge you have of appropriate techniques to get things done.

Recognizing both components of politics and their importance is critical for project managers. For example, politically naive project managers depend too much on the tangible component of politics and believe knowing policies and procedures is sufficient. They believe they can manage their projects by going by the book only and by using a formal approach in communicating and dealing with stakeholders. They believe politics are bad and unpleasant, and therefore, they ignore or avoid politics.

Experienced and politically savvy project managers recognize politics are a necessary part of life in project environments, and they are important to get things done through others. They believe you should develop an effective network and know the right people in the right places to achieve desired outcomes. To increase their network power, they must nurture and feed the network and use this power ethically, fairly, and carefully. For example, they must make deposits first in this network account (by doing favors for others); otherwise, their checks bounce when they ask for favors and help.

Figure 1.2 shows that power combined with politics results in successful project management. Project managers can increase their informal power by integrating their knowledge, skills, and experience. They can also become more politically savvy by understanding and using both tangible and intangible components of politics. Successful project

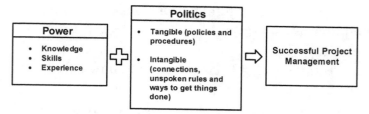

management is achieved by a better understanding of both power and politics and by using them effectively in dealing with project stakeholders.

Figure 1.2. Power and politics in project management.

Project managers must understand the dynamics of politics to meet challenges when dealing with stakeholders. Politics have two components: 1) tangible and obvious and 2) intangible and unobvious.

Politically naive project managers depend too much on the tangible component of politics and believe knowing policies and procedures is sufficient. Power combined with politics results in successful project management. Successful project management is achieved by a better understanding of both power and politics and by using them effectively to deal with project stakeholders.

1.4. Politics Are About Power

> *Politics are about gaining power and*
> *then using it to get what you want.*
> —Vijay Verma

Power is necessary to manage politics in any organization. Project managers who recognize the power they have and know how to use it effectively manage projects better. Therefore, project managers must increase their total power composed of formal and informal power.

Total Power = Formal Power + Informal Power

Formal power refers to power granted according to hierarchical position in the organizational chart, whereas *informal power* refers to what project managers earn based on their knowledge and experience. Because project managers do not usually have much formal power based on their position in the organizational chart, they must strive to increase the informal power granted them because of their knowledge, experience, interpersonal skills, and leadership skills. As a step toward gaining power and learning to manage politics better, project managers must remember the following three steps:

1. Seeking power

2. Acquiring power

3. Maintaining power by using it effectively

1.4.1. Seeking power

There is no knowledge that is not power.
—Ralph Waldo Emerson

Power is the key to managing politics effectively, but you must understand which power (formal or informal) is more useful. Project managers must recognize that formal, or positional, power can only be increased by going up the management ladder, and it can be taken away when they move from that position.

Informal power is more essential and useful in influencing stakeholders and managing politics. Therefore, project managers must recognize the need to increase the informal component of total power because they can work to increase their informal power whenever they want and on their own. The use of formal power leads to command and control, whereas informal power helps project managers influence their stakeholders positively to complete project assignments efficiently and deliver successful projects.

Many people think formal and positional authority (determined by the status in the organization chart) is power. However, an in-depth analysis suggests that there is more to power than just formal authority. The following are the three ingredients of power (Ferrarro 2010):[2]

1. Knowledge

2. Skills

3. Experience

These ingredients are incorporated in informal sources of power: referent, expert, information, network, and persuasion power. Using power effectively is the key to successful outcomes. Knowledge is power only when it integrates with experience and skills, followed by appropriate action. All three ingredients must tightly integrate in managing projects successfully. If any of these ingredients is missing or unbalanced, the overall skill set is weakened, thereby increasing risks of project failure; for example (Ferrarro 2010):[3]

- Knowledge without experience leads to lack of confidence.

- Skills without knowledge and experience lead to not getting traction on your project or doing the wrong things. It also might make project managers over dependent on processes and tools.

- Knowledge and experience without skills and actions lead to just talking and no results. Such leaders keep telling "what" should be done, but they do not know "how to." Therefore, they cannot direct others how to use the knowledge and experience effectively, which makes them poor practical leaders.

- Experience without knowledge and skills leads to hollow confidence. It leads to the illusion that people can do things or get things done, but when doing it, they do not know "what" to do and "how" to get desired results.

The enhancement of each ingredient of power should be viewed holistically because each ingredient is mutually exclusive and complements the others. The need to seek power should be an objective to navigate through organizational politics. The eight sources of power are shown in table 1.1. Five out of eight are informal powers, with the remaining three being formal powers. If you have legitimate, or positional, power, you typically have reward and coercive powers. Therefore, you should work to gain informal powers because they make up a significant part of the overall power package, and informal powers are more permanent.

	Sources of Power	Type	Based On
1	Legitimate	Formal	• Positional power • Access to resources (money and people)
2	Reward	Formal	• Distribution of rewards • Developing effective rewards
3	Coercive	Formal	• Poor performance review with a plan to improve • Removal of rewards
4	Referent	Informal	• Identification with high-profile people and projects • Leadership skills
5	Expert	Informal	• Variety of knowledge and experience • Degree of confidence of management
6	Information	Informal	• Knowledge of how organization works • Important information link
7	Network	Informal	• Personal and business contacts • Doing and receiving favors
8	Persuasion	Informal	• Winning others' cooperation (negotiation) • Ability to enhance trust and agreement

Table 1.1. Eight Sources of Power (Types and Bases)

1.4.2. Acquiring power

Thoughts mixed with definiteness of purpose, persistence, and a burning

> *desire are powerful things.*
> —Napoleon Hill

After project managers decide to seek more power, the next step is to acquire it. Project managers must develop strategies to increase their power, with special emphasis on increasing their informal powers because of the following four advantages:

1. No one can take away informal powers. By comparison, formal powers (legitimate, reward, and coercive powers) belong to a person shown inside a particular box in the organizational chart, and you lose these powers when you move from that position in the box or when you leave your particular position, department, or organization.

2. Project managers can immediately work to acquire informal powers. They can work to increase their informal powers (referent, expert, information, network, and persuasion powers) by studying on their own and getting help from their mentors without waiting for approval or promotion.

3. The more you share informal powers, the more you get. Project managers can increase their informal powers even more when they share them with other project personnel. For example, the more they mentor someone about subject matter or leadership skills, the more they learn. You learn much by teaching others, and you develop more self-confidence and increase others' confidence. Similarly, network power increases when people share

their network with others, for example, through LinkedIn and other such systems.

4. A stronger foundation of informal powers leads to better opportunities for formal power. For example, if project managers have sufficient informal powers (referent, expert, information, network, and persuasion powers), senior management thinks of them for promotions and new opportunities. In addition, when project managers get a promotion based on their informal powers, others are happy for them and believe they deserve that promotion, which leads to better acceptance of the promotion and, hence, more cooperation and collaboration from their colleagues, team members, and other stakeholders.

1.4.3. Maintaining power

The key to maintaining power is to use your power informally.
—Vijay Verma

Maintaining power refers to how you use your power critical to managing politics effectively. Having power is important, but knowing how to use it effectively is even more important to maintain your power. Maintaining power also helps project managers influence their stakeholders continually because

Influencing = Having Power + Exercising Power

Project managers lose their power and cannot maintain it if they use it inappropriately. Usually, the use of power informally is better. However, when necessary, project managers should not hesitate to use their powers formally, depending on the situation, but usually as a last resort. This strategy increases their political savviness because they make proper decisions and manage their people and projects with the appropriate skills to suit the situation. Project managers cannot manage politics

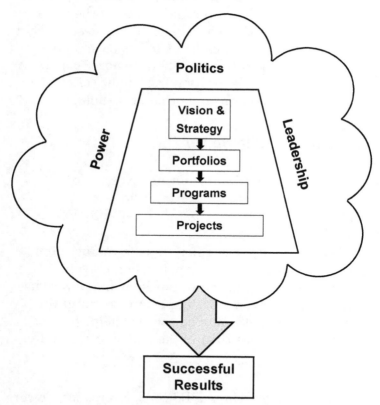

Figure 1.3. From strategies to results with power, politics, and leadership.

60

effectively if they do not maintain their power, giving them the ability to influence their stakeholders to get things done.

As shown in figure 1.3, senior management must develop vision and strategies and break them into appropriate portfolios, programs, and projects. They should emphasize to their managers the importance of understanding the dynamics of power, politics, and leadership, which are complex and unpredictable, to achieve successful projects.

Power and politics are interrelated in that you need power and you need to use it appropriately to navigate politics effectively. You must seek power with a proper balance of knowledge, skills, and experience. A project manager's total power is made up of both formal and informal power. Project managers should acquire power through legitimate means and focus on increasing their informal powers because of four distinct advantages. To manage politics effectively, project managers must be able to maintain their power by using it informally.

Chapter 1 Summary

Politics are complex because influencing skills for managing conflicts are individualized, varying from person to person. Conflict management is an individualized that involves persuading and influencing others to reach a common agreement based on standards of fairness and priority for maintaining good working relationships. Often, when people have very strong and different viewpoints and expectations and they are not open enough to discuss their differences with others to resolve conflicts and find common ground, they use politics to achieve their objectives and interests in their way. This attitude leads to negative politics, whereby people try to gain control rather than collaborating and working as a team to meet project objectives.

Usually, project managers have great responsibility, but no direct formal authority. Power enables them to influence stakeholders. Intensity of politics among stakeholders consists of four variables:

1. Degree of interaction

2. Frequency of interaction

3. Competing stakeholder requirements and agenda

4. Stakeholders' desire to reach an agreement

The intensity of politics project managers face increases as they manage projects, programs, and then portfolios, and as more people with diverse interests from different departments are involved and expected to work in teams.

Project managers must understand the dynamics of politics to meet challenges when dealing with stakeholders. Politics have two components:

1. Tangible and obvious

2. Intangible and unobvious

Politically naive project managers depend too much on the tangible component of politics and believe knowing policies and procedures is sufficient. Successful project management is achieved by a better understanding of both power and politics and by using them effectively in dealing with project stakeholders.

Project managers must increase their total power composed of formal and informal power.

Total Power = Formal Power + Informal Power

Formal power refers to power granted according to position in the organizational chart, whereas *informal power* refers to what project managers earn based on their knowledge and experience. Project managers must recognize the need to increase the informal component of power. Formal power leads to command and control, whereas informal power helps project managers influence their stakeholders positively to complete project assignments efficiently and deliver successful projects. Power has three ingredients:

1. Knowledge
2. Skills
3. Experience

There are eight sources of power:

1. Legitimate (formal)
2. Reward (formal)
3. Coercive (formal)
4. Referent (informal)
5. Expert power (informal)
6. Information power (informal)

7. Network power (informal)

8. Persuasion power (informal)

To manage politics, project managers must remember the three steps of seeking power, acquiring power, and maintaining power. You should gain informal powers because they are a significant part of the overall power package and more permanent. After seeking power, project managers must acquire and especially increase their informal powers, which have four advantages:

1. No one can take away informal powers.

2. Project managers can immediately work to acquire informal powers.

3. The more you share informal powers, the more you get.

4. A stronger foundation of informal powers leads to better opportunities for formal power.

To manage politics effectively, project managers should emphasize maintaining power. *Maintaining power* refers to how you use your power critical to managing politics effectively. Having power is important, but knowing how to use it effectively is even more important. Maintaining power also helps project managers influence their stakeholders continually because

Influencing = Having Power + Exercising Power

Politics in Project Management

Politics are inevitable. People will disappear, but not the politics.

—Vijay Verma

This chapter covers how the intensity of politics varies throughout the project life cycle. Effective project managers observe project dynamics, and they can see whether and when the project gets in trouble. They use their power and political skills to survive turbulent phases. A troubled project's phases are described in terms of how stakeholders respond to these phases emotionally to cope with the circumstances and how they could approach proactively to manage the six phases. Practical tips are presented to help project managers navigate all six phases to minimize troubled projects' negative impact on the organization.

2.1. Project Life Span (PLS) and Politics

*Learn to manage politics; otherwise, you
will become a victim of them.*

—Vijay Verma

The intensity of politics increases as more and
more people and departments are involved in the
overall project management process. It varies as you
move from project manager to program manager and
then to portfolio manager because of diverse stakeholders from various departments, business units,
and management levels. This variation is because
people in each area of the organization have their
set way to do things, and they might be inflexible.
Politics are unpredictable, and the intensity of politics varies as the project progresses from conceptual
phase to detail, execution, and finishing phases.

Projects have a distinct life span that starts
with an idea or a new concept about a product or
service in the conceptual phase (C). They then
progress through phases of detailed planning (D),
execution (E), and finishing (F) when the outcome or
deliverable (new or improved product or service) is
transferred to the project owner, customer, or client.
This sequence is commonly known as the *project life
cycle (PLC)*.

The projects' size and nature and organizations'
business and industrial environments influence
the detailed descriptions of PLC phases. In simple
terms, the four generic phases can be combined in
two phases:

1. Plan (includes conceptual (C) and detailed planning (D))
2. Produce (includes execution (E) and finish (F) phases)

It can be argued that this is not a PLC because these stages can occur several times throughout one project. There might be repetition; for instance, the developed plan's execution might lead to replanning after discovering unknown items. At the same time, there is often overlap between PLC phases.

From a timeline viewpoint, *project life span (PLS)* might be more appropriate than PLC because projects take a certain time as they progress through the concept phase to the finishing phase. Different activities occur during each PLS phase. These activities lead to many issues with processes, people, and politics.

Figure 2.1 shows the four PLS phases and the levels of politics and the change in levels of politics and personnel according to the project life span. The level of politics is high in the beginning, and it goes down a bit once the front-end issues are addressed, but it goes up again as plans are executed, and more people and personalities are involved. Good leaders address these issues continually to minimize the negative impact of politics.

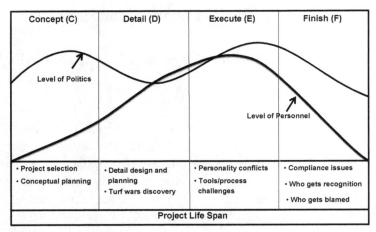

Figure 2.1. Level of politics according to project life span.

In a project environment, politics influence the following:

- Project selection, review, and management
- Dynamics among team members
- Senior management's view of project benefits for overall organizational strategy

Furthermore, project politics have four dimensions:

1. Issues related directly to the project and its governance (priorities, resource allocation, and so on)

2. Team environment (interactions, conflicts, rewards, and so on)

3. Overall organizational environment (strategies versus resource constraints for dollars, people, and so on)

4. Cultural factors (internal and external), regulations, and so on

The intensity of politics increases as more and more people and departments are involved in the project management process. Politics are unpredictable, and the levels of politics vary as the project progresses through its phases. The projects' size and nature and organizations' business and industrial environments influence the detailed descriptions of PLC phases. From a timeline viewpoint, project life span (PLS) might be more appropriate than PLC because projects take a certain time as they progress through the concept phase to the finishing phase.

External and internal politics affect overall project progress and outcome. External politics include politics in client organizations, and actions and demands that external stakeholders, such as regulatory authorities and different levels of government, enforce. Internal politics include power struggles among various stakeholders, project personnel's political behaviors, the prioritization system and allocation of resources, and project monitoring (gate reviews and so on).

Project champions can help navigate internal politics by using their power and influence in the organization, and external politics by using their network and persuasion power. Important factors are the depth and breadth of a project manager's knowledge, skills, and experience in managing projects and level of understanding about the organization's overall political dynamics, which includes understanding the big picture and the workings of the organization.

2.2. Dynamics of a Troubled Project (the Emotional Side)

> *After two weeks of working on a project, you know whether it will work or not.*
> —Bill Budge

Most projects typically go through four generic phases:

1. Conceptual (C)
2. Development (D)
3. Execution (E)
4. Finishing (F)

However, projects lacking appropriate planning, resources with proper skill mix, and leadership get in trouble. In such cases, many stakeholders play political games to protect themselves. This section describes the six typical phases of a troubled project and how stakeholders often respond to those phases to survive and deal with the dynamics of project personnel and related politics emotionally (negatively) and proactively (positively) (Verma 2008–2015).[1]

Table 2.1 shows the emotional view of six phases driven by project dynamics and associated politics.

Phases 1 and 2 are typical for most projects, and phase 3 responds to lack of proper planning, poor management, and ineffective leadership, which happens when a team realizes the project scope is much larger and more complex than originally anticipated or planned. After phases 1 and 2, project

	Phase	Emotional View	Driven By
1	Celebration	Enthusiasm	Project Life Cycle and Project Dynamics
2	Disillusionment	Confusion/Worry	
3	Panic	Stress	
4	Search for the Guilty	Finger-Pointing	Project Politics
5	Punishment of the Innocent	Unfair Treatment	
6	Praise and Honor for Nonparticipants	Letdown	

Table 2.1. Emotional Views of a Troubled Project

managers or people in charge of projects realize they must hurry to work on the project. The natural first step before starting a project should be to do good front-end planning and prepare a detailed plan. However, in some cases, people start working on work packages, project tasks, and activities without proper planning, which triggers a series of problems.

People work in panic because they think they must do something immediately; otherwise, they could be significantly behind schedule. Panic happens when project schedules are unrealistic, and appropriate stakeholders are not involved in overall planning. People work without thorough analysis and evaluation of project risks, resource constraints, and available options.

The following are common consequences of working in panic:

- Rework and extra work

- Shortcuts to finish project activities

- Wrong techniques and poor quality deliverables

- Irrational decisions and conflicts
- Stressful working environment

When panic starts and the consequences occur, phases 4–6 likely follow. Panic is like a storm, and experienced project managers should observe the signals and prepare themselves for the storm that includes phases 4–6.

As shown in table 2.1, project life cycle and project dynamics drive phases 1–3, whereas project politics drive phases 4–6. Phases 4–6 occur when the project gets in trouble, and people from management want to protect themselves from the negative consequences of project outcomes. They deflect blame to others and play political games to make themselves look good, even at the cost of others. Often, they blame the project manager and team members for things that went wrong, even if their poor leadership caused the mistakes. Sometimes, even sponsors play such political games. At the same time, they take credit for the well-managed part of the project and the right things they intended to do or start, but which they never followed or completed successfully.

Project managers must evaluate the overall project environment and address the following question: What should you do when the project gets in trouble because of panic and then goes through phases 4–6?

Here, effective project managers must understand the dynamics of power and politics, which are this book's focus. Politically savvy project managers can protect themselves and help their team through these rough phases. They should anticipate future problems and plan actions they, the sponsor, and other stakeholders should take to survive these emotional phases.

2.2.1. Phases driven by project life cycle and project dynamics

Most projects are derailed by negative politics despite good planning.

—Vijay Verma

Phase 1—Celebration

Enthusiasm is important and natural. However, you must keep it within limits. This phase's purpose is to celebrate team success in winning the bid.

Phase 2—Disillusionment

Project managers and key team members might feel disillusioned and confused about schedules, budgets, outcomes, and deliverables promised in the bid document, because they are too optimistic. It happens because of lack of supporting details before finalizing the bid document and crosschecking the plan thoroughly.

Phase 3—Panic

Panic leads to stress, but it happens only if there is no overall front-end planning and creative leadership. You must analyze and plan all work packages in detail. Work-package managers should have identified all project risks and developed high-level strategies and detailed plans to complete project activities without crisis. These actions help avoid panic and increase the confidence of the project sponsor, project manager, and team members to deal with the main issues and to prepare for unanticipated problems.

2.2.2. Phases driven by project politics

"When you blame and criticize others, you are avoiding some truth about yourself."
—Deepak Chopra

Phase 4—Search for the guilty

During this phase, many people are nervous and frightened; they feel vulnerable and start pointing fingers at others to protect themselves. Project managers should keep their heads low (keep a low profile) and not become too vocal at sensitive meetings, especially when the "search for the guilty" is happening. It is better not to take responsibility to resolve things that are badly broken, too political, or beyond your control.

Phase 5—Punishment of the innocent

Who are the "innocent" people who are punished? Usually, in troubled projects, project managers and hardworking team members are punished, which is unfair. To avoid punishment, project managers should anticipate the possibilities of finger-pointing behavior and potential restructuring of the organization. They should keep their eyes and ears open and network more with people in power rather than spending all their time working on project processes, schedules, meeting minutes, and so on. They should understand and address the main issues and concerns of senior management and use this understanding to protect themselves and their hardworking team members.

Phase 6—Praise and honor for nonparticipants

This phase recognizes people who never contributed significantly with praise and honor. Aggressive and ambitious senior management members want praise and a positive image so they can move up in the organization. Sometimes, they steal credit from others to meet their objectives. Project managers then feel let down by these actions. They also want to be included in all-important organizational publicity photos.

Politically savvy project managers should ensure senior management remembers them for their important contributions to the project, gives them proper recognition, and includes them in those important photos. To achieve this, project managers should ensure "they are seen, and not forgotten," which implies senior management should not forget the project managers' hard work and their help in resolving many project challenges and problems.

When management recognizes project managers with good people skills, and they are included in those photos, they should also include their key team leaders in those photo opportunities. However, if those team leaders cannot be included in the pictures for logistical reasons, project managers should recognize them at other suitable forums to positively reinforce them. This recognition helps project managers gain team members' trust, loyalty, and long-term cooperation. Suggestions to survive phases 4–6, which are more political, are summarized in table 2.2.

Political Phases	Suggestions to Survive
Search for the Guilty	Keep your head low and be cautious.
Punishment of the Innocent	Network more, and understand dynamics of power (to identify movers and shakers).
Praise and Honor for Nonparticipants	Be seen and not forgotten (to be recognized and included in promotional pictures with senior management).

Table 2.2. Survival Suggestions for Political Phases

There are six phases of a troubled project: 1) celebration, 2) disillusionment, 3) panic, 4) search for the guilty, 5) punishment of the innocent, and 6) praise and honor for nonparticipants. From an emotional viewpoint, phases 1–3 (enthusiasm, confusion/worry, stress) are driven by project lifecycle and project dynamics, and phases 4–6 (finger-pointing, unfair treatment, letdown) are driven by politics.

People feel happy and enthusiastic when they celebrate any success. When projects are not well planned, and project goals are unrealistic, project teams are often disillusioned and confused. Project teams then typically panic and start the project immediately without proper plans, leading to stress for everyone on the team. When people work under panic and stress, things go wrong, quality goes down, and people point fingers at one another to deflect accountability. After searching for the guilty (finger-pointing), hard-working project managers and team members typically are unfairly punished because of their weakness in navigating politics. Unfortunately, senior management who did not fulfill their responsibilities still receive the praise and honor they do not deserve, which makes project managers and team members unhappy.

Project managers must anticipate when projects are on the verge of trouble and stay calm and avoid acting emotionally. Instead, they should approach the projects proactively to get them back on track and keep team spirits high.

2.3. Dynamics of a Troubled Project (the Proactive Side)

> *I don't work on a project unless I believe that it will dramatically improve life for a bunch of people.*
> —Dean Karnen

Whenever communication, trust, and teamwork are lacking, overall performance and morale of project personnel go down, and projects get into trouble, leading to significant delays in schedule and cost overruns. Then, people become emotional and react to circumstances rather than pausing to analyze overall project dynamics objectively before taking major actions.

Experienced project managers do not become too emotional and react negatively because it leads to negative politics and poor working relationships with team members and other project stakeholders. They focus on overall analysis of the big picture, drill to understand the real issues and problems, and act proactively to look at a troubled project's phases more positively rather than with an emotional (negative) view. This approach redirects people's energy from a negative side to more collaboration, teamwork, and commitment, which helps identify the real issues, problems, and risks and encourages project personnel to work together to find better options and solutions to put the project back on the right track.

Table 2.3 shows the proactive (positive) views of a troubled project's six phases with focus on observing each phase carefully and becoming more proactive. Project managers should analyze the situation and stakeholders' reactions thoroughly during each phase. They should then prepare a plan to best cope with the circumstances during respective phases to avoid future problems, especially for phases 4–6 driven by politics.

	Phase	Proactive View	Driven By
1	Celebration	Kickoff (initiation)	Project Manager and Project Sponsor
2	Disillusionment	Reality Check (analysis of plans and deliverables)	
3	Panic	Cautious Fast Track (with risk management)	
4	Search for the Guilty	Status Reviews and Post-Mortem Analysis (what went wrong and why)	Project Management Office
5	Punishment of the Innocent	Evaluation of Performance (use objective criteria!)	
6	Praise and Honor for Nonparticipants	Recognition of Right People (avoids favoritism)	

Table 2.3. Proactive Views of a Troubled Project's Phases

2.3.1. Phases driven by project manager and project sponsor

> *"One of the true tests of leadership is the ability to recognize a problem before it becomes an emergency."*
> —Arnold Glasgow

Phase 1—Celebration

The celebration phase can be used as an opportunity to have a good kick-off meeting and start the project on the right foot. In this phase, senior management and the sponsor announce the "go ahead" decision and initiate the project. Critical success factors should be defined by how the project outcome will be evaluated and measured. The project manager and key team leaders are named with project scope, major milestones, management expectations, and the project's overall importance to the organization. You must define expectations from the project team and other stakeholders and clarify them by seeking feedback, helping minimize surprises and get buy-in from stakeholders.

Phase 2—Disillusionment

This phase should focus on a reality check rather than feeling disillusioned because of a slightly optimistic schedule, budget, and deliverables promised in the bid document. In this phase, project managers should analyze project scope, budget, schedule, and deliverables with team members. They should review the details used to create the overall budget and schedule and identify risks and plans to mitigate risks and minimize their negative impact on the overall project. You must review sponsor and client expectations and acquire special resources, if needed, to meet project objectives within time and budget constraints to avoid stress and panic.

Phase 3—Panic

A cautious fast track with risk management helps avoid panic situations. The panic stage involves scope management planning and often reveals that the budgets and schedules are tight, and fast tracking might be needed with proper risk management strategies.

Table 2.4 shows the advantages of proactivity in political phases. The proactive view emphasizes identifying risks and mitigating strategies.

Political Phases	Advantages of Proactivity
Search for the Guilty	Regular status reviews to identify risks, warnings, and possible solutions
Punishment of the Innocent	Keeps the focus on problems and issues, rather than on people
Praise and Honor for Nonparticipants	Recognition of the right people identifies the most valuable performers, regardless of their positional power

Table 2.4. Advantages of Proactivity for Political Phases

Sometimes, difficult decisions must be made quickly to meet deadlines, even though some view it as working in panic mode. Nevertheless, good project managers do risk analysis, monitor project progress continuously, and take corrective actions to resolve significant variances, as needed. Project managers should keep sponsors informed of requirements for additional resources and major challenges and issues (known and unknown). You must review them continuously when controlling changes to scope.

2.3.2. Phases driven by project management office

> *Power is no blessing in itself, except when*
> *it is used to protect the innocent.*
> —Jonathan Swift

Phase 4—Search for the guilty

This phase focuses on identifying and evaluating real problems objectively with stakeholders and helps minimize finger-pointing. This phase aims to determine what went wrong, why it happened, and how it can be avoided in the future. You must conduct reviews objectively and honestly. Do not discourage people from presenting their honest feedback and analysis. The purpose is to encourage people to learn from their mistakes and build organizational assets rather than to punish them for making those mistakes. Regular status reviews will identify problems and potential solutions. This approach empowers people and makes them feel free and comfortable in presenting honest data and developing innovative ideas to resolve problems.

Phase 5—Punishment of the innocent

The aim in this phase is to translate failures into lessons learned by emphasizing issues and problems rather than pointing fingers at people. Separating people from problems and evaluating these problems objectively does not put people on the spot and, therefore, leads to constructive discussions and a better outcome. It adds more objectivity in the overall project management process and avoids punishing the wrong people, encouraging people to think outside the box and develop creative solutions.

Phase 6—Praise and honor for nonparticipants

In this phase, high achievers should be recognized so they feel positively reinforced and more self-motivated in the end, but without favoritism. Objective and fair criteria should be used to recognize people regardless their position of power. They should be encouraged for their efforts and assured of continuous management support.

Project activities at the beginning of the project life cycle drive phases 1–3 in a troubled project. Better front-end planning, scope management, and positive actions for phases 1–3 should help minimize the negative impacts of these phases.

Politics drive phases 4–6 in an emotional view. Therefore, to prevent the negative impact of these phases, project managers must anticipate political implications that lead to such emotions and, hence, negative politics. This anticipation allows them to achieve more positive and practical outcomes and deal with these phases positively. Setting up a Project Management Office (PMO) fully supported by senior management and led by experienced project management practitioners with good people skills and the ability to relate to the "big picture" helps minimize the negative impacts of politics and unnecessary conflicts during phases 4–6.

Project managers should develop creative leadership and understand organizational dynamics, the project environment, and working relationships among various project stakeholders and team members in their power levels and political behaviors.

Good project managers approach troubled project phases proactively to defuse negative emotional views. There are six phases of a troubled project: 1) celebration, 2) disillusionment, 3) panic, 4) search for the guilty, 5) punishment of the innocent, and 6) praise and honor for nonparticipants. From a proactive viewpoint, phases 1–3 (kickoff, reality check, cautious fast track) are driven by the project manager and project sponsor, and phases 4-6 (status reviews and post-mortem analysis, evaluation of performance, recognition of right people) are driven by the PMO.

A kickoff, or initiation, is a typical proactive approach to celebration. When projects are not well planned, project teams are often disillusioned and confused about the project plans and deliverable timelines. The proactive project manager does a reality check and analyzes plans and deliverables. If panic sets in, the proactive project manager proceeds on a cautious fast track with risk management. When people work under panic and stress, things go wrong, quality goes down, and people point fingers at one another to deflect accountability. A proactive approach at this time would be to perform status reviews and post-mortem analysis to determine what went wrong and why. After searching for the guilty (finger-pointing), hard-working project managers and team members typically are unfairly punished because of their ineptitude in navigating politics. A proactive approach would evaluate performance with objective criteria. Unfortunately, senior management who did not fulfill their responsibilities still receive the praise and honor, which lets down project managers and team members. Project managers should avoid favoritism to recognize the right people.

Instead of responding emotionally when projects become troubled, project managers should approach the projects proactively to get them back on track and keep the team motivated.

Chapter 2 Summary

Most projects go through four phases:

1. Conceptual (C)
2. Development (D)
3. Execution (E)
4. Finishing (F)

Projects lacking appropriate planning, resources with the proper skill mixture, and leadership get in trouble. A troubled project has six phases:

1. Celebration
2. Disillusionment
3. Panic
4. Search for the Guilty
5. Punishment of the Innocent
6. Praise and Honor for Nonparticipants

During the celebration phase of a troubled project, people feel happy and enthusiastic from

84

an emotional viewpoint. A proactive approach to celebrating would be a kickoff, or initiation. When projects are not well planned, and project goals are unrealistic, project teams are often disillusioned and confused from an emotional viewpoint, which can be handled proactively by a reality check with the project manager analyzing plans and deliverables. From an emotional viewpoint, project teams then typically panic and start the project immediately without proper plans, leading to stress for everyone on the team. During this phase, the proactive project manager proceeds on a cautious fast track while managing risk.

When people work under panic and stress, things go wrong, quality goes down, and people point fingers at one another to deflect accountability. A proactive approach at this time would be to perform status reviews and post-mortem analysis to determine what went wrong and why. After searching for the guilty (finger-pointing), hard-working project managers and team members typically are unfairly punished because of their weakness in navigating politics. A proactive approach would evaluate performance with objective criteria. Unfortunately, senior management who did not fulfill their responsibilities still receive the praise and honor they do not deserve, which makes project managers and team members unhappy. Project managers should avoid favoritism to recognize the right people.

Phases 1 and 2 are typical for most projects, and phase 3 responds to improper planning, poor management, and ineffective leadership. Phases 4–6 occur when the project gets in trouble, and people from management want to protect themselves from the negative consequences of project outcomes.

From an emotional viewpoint, the project lifecycle and project dynamics drive phases 1–3, and politics drive phases 4–6. Survival suggestions are offered for phases 4–6. From a proactive viewpoint, the project manager and project sponsor drive phases 1–3, and the project management office (PMO) handles phases 4–6. This chapter describes various actions senior management, project managers, and the PMO can take to minimize the negative impact of troubled projects and put them back on the right track.

Politics are done by people, and wherever there are people, there are politics. The intensity of politics increases as more and more people and departments are involved in the project management process. Politics are unpredictable, and the level of politics changes as the project moves through its phases. The detailed description of phases of a project life cycle (PLC) differs depending on project size and industry. Project life span (PLS) might be more appropriate from a timeline viewpoint than PLC because projects take a certain time as they progress through the concept phase to the finishing phase.

In a project environment, politics influence the following:

- Project selection, review, and management
- Dynamics among team members
- Senior management's view of project benefits for overall organizational strategy

86

Furthermore, project politics have four dimensions:

1. Issues related directly to the project and its governance (priorities, resource allocation, and so on)

2. Team environment (interactions, conflicts, rewards, and so on)

3. Overall organizational environment (strategies versus resource constraints for dollars, people, and so on)

4. Cultural factors (internal and external), regulations, and so on

Internal politics include power struggles among various stakeholders, project personnel's political behaviors, the prioritization system and allocation of resources, and project monitoring. Project champions can help navigate internal politics by using their power and influence in the organization, and external politics by using their network and persuasion power.

Whenever communication, trust, and teamwork are lacking, overall performance and morale of project personnel go down, and projects get into trouble, leading to significant delays in schedule and cost overruns. Experienced project managers do not become too emotional and respond negatively because it leads to negative politics and poor working relationships with team members and other project stakeholders.

Project managers should approach a troubled project's six phases positively with focus on observing each phase carefully and proactively. The proactive view emphasizes identifying risks and mitigating strategies.

Part I Summary

Politics are a necessary part of project life because projects are done by people with a diverse mixture of backgrounds, skills, and experiences. People often assume politics do not and should not exist, and therefore, they do not need to learn any skills to understand and manage politics effectively. Most project managers are good in planning and executing project activities to meet organizational goals, but are generally poor in navigating organizational politics.

In any organization, you must understand the importance and dynamics of politics, develop skills to manage politics effectively, and learn to become politically sensible. Good project managers recognize that there are various shareholders with different interests, expectations, and personalities throughout the organization. Because many stakeholders have different viewpoints about politics and prefer their way of doing things, a better understanding of politics and stakeholder management are often necessary to get things done to deliver successful projects.

The concepts of power and politics are important in project management, but many project managers do not fully appreciate their impacts. Project managers have enormous responsibility but limited formal authority over their project stakeholders. Many people think of power as formal authority. However, there is an informal component that can only be earned by increasing knowledge and experience. Project managers should gain informal powers because these powers are more permanent, and they represent a significant component of the overall power package. Furthermore, they should use their powers informally to maintain them.

The dynamics of politics are complex. Power and politics are interrelated in that project managers must have significant power to navigate organizational politics. To manage politics effectively, project managers should seek power, acquire power, and maintain power by using it appropriately.

Project managers should also understand the tangible component of politics (driven by organizational policies and procedures) and intangible component of politics (driven by connections and unspoken ways of getting things done) and learn to manage politics at all organizational levels. The intensity of politics varies as a project progresses from the conceptual stage to the detailed, execution, and finishing stages. The intensity of politics varies, also, as project managers start managing programs, and then portfolios of programs, because more people with diverse interests from different departments are involved and need to work together.

A troubled project has dynamics that can be explained by the six phases of a troubled project. These six phases have an emotional view and a proactive view that can be considered negative and positive, respectively. Project managers should understand the dynamics of a troubled project and take various actions to minimize the negative impacts of consequences and put the troubled project back on the right track.

Power and politics are interrelated. Project managers should mostly use their power informally to gain cooperation and commitment. They should use power formally only as a last resort because it typically leads to compliance, not commitment.

Politics are part of life in project environments. Many projects are derailed because of politics despite good planning and project management processes. Project managers must understand the dynamics of politics and their use to get things done from project stakeholders. Politics have a negative connotation; however, politics can also be positive.

Part II. Dynamics of Politics

Chapter 3: Two Types of Politics

3.1. Negative Politics

3.2. Positive Politics

Chapter 3 Summary

Chapter 4: Why Organizational Politics?

4.1. Projects and Project Management Issues

4.2. People Behavioral Issues

4.3. Organizational and Management Issues

4.4. External and Global Market Environment Issues

Chapter 4 Summary

Chapter 5: Politics in Managing Stakeholders

5.1. Politics and Stakeholders

5.2. Analyzing Stakeholders

5.3. Prioritizing Stakeholders Using Main Attributes

Chapter 5 Summary

Part II Summary

Dynamics of Politics

Man is by nature a political animal.
—Aristotle

Part II describes the attributes behind negative politics and positive politics and emphasizes how positive politics can rechannel people's energy from whining and propagating negative energy toward creating a culture of more collaboration, teamwork, and cooperation. Eliminating politics is almost impossible, but you can minimize their undesirable impact by using key ideas associated with positive politics.

Politics occur in most organizations for many reasons. This part describes various reasons and circumstances that create politics, including the four main issues that lead to politics in most organizations:

1. Project and project management issues
2. People behavioral issues
3. Organizational and management issues
4. External and global market environment issues

Stakeholders push their ideas and processes to manage projects because they have diverse beliefs, values, and expectations, which lead to politics. Project managers must recognize the importance of politics to manage their stakeholders. They must understand their internal and external stakeholders and analyze their requirements based on various attributes.

Part II deals with stakeholder analysis and prioritizing them by what they want and why and covers various ideas to manage stakeholders with different interests and abilities effectively. It deals with the politics of managing stakeholders and their interests in a politically perceptive manner.

Two Types of Politics

*Diplomacy is an art of letting people
have things your way.*

—Italian diplomat

Politics are inevitable in most organizations, especially in project environments, because people do most projects, and wherever there are people, there are politics. Most project managers view politics negatively. However, politics can be positive. This chapter describes the dynamics of positive politics, emphasizing that positive politics can change the organizational culture and create a climate of improved cooperation, collaboration, and teamwork. We must understand the dynamics of politics and project stakeholders' political behaviors so we can learn to manage projects effectively and not become political victims.

Pinto (1994, 111–130) described two views of politics: 1) a negative view and 2) a neutral view.[1] A negative view of politics suggests politics are bad and wasted time because people use politics to satisfy self-interests, often at the cost of meeting overall organizational goals. A neutral view of politics argues that politics and political activities are a natural part of organizations and similar to an organizational culture and organizational structure. The politics, as such, are neither good nor bad until they are used. A neutral way suggests politics are essential to project management, and we must learn and apply them unselfishly for the organization's benefit (Beeman & Sharkey 1987, 2, 26–30; Markus 1983, 321–342; Allen et al. 1979, 1, 78–93).[2]

Politics can be positive and negative. Negative politics focus on self-interests rather than the organization's objectives. Positive politics can be described as an approach that emphasizes the dynamics of politics and suggests using effective management and leadership techniques to enhance open communication, challenge the processes, gain trust and commitment, and encourage information sharing. Such techniques create a culture of more collaboration, cooperation, and synergy to achieve extraordinary results.

This chapter describes both types of politics, the reasons they happen, and the main events leading to negative and positive politics. I illustrate how project managers can use certain techniques and the main themes behind positive politics to redirect people's energy from a negative direction that emphasizes command and control and hidden agendas to a positive direction that emphasizes collaboration, co-operation, trust, and teamwork. Attributes and ideas create negative politics as well as positive politics. These are described in the following section.

3.1. Negative Politics

> *No matter how hard you work for success, if*
> *your thought is saturated with fear or*
> *failure, it will kill your efforts, neutralize*
> *your endeavors, and make success impossible.*
> —Baudhuin

Negative politics emphasize "I." The main theme of negative politics is to "divide and conquer." People use negative politics mostly to fulfill self-interests, rather than to benefit the organization. Negative politics are common in many organizations, and they are generally predatory. Project managers should not ignore politics; otherwise, they become a political victim. Instead, they should learn to minimize the negative impacts of politics on projects and organizations. Negative politics create frustration, anger, anxiety, and lack of trust in people and processes.

Negative politics involve activities to acquire, develop, and use power to get your way (Pfeffer 1981, 7).[3] According to Henry Mintzberg (1983, 421), a well-known researcher in organization theory, politics encompass modern organizations, which he refers to as *systems captured by conflict.*[4] Another definition of politics suggests that political behavior is highly competitive, and it focuses on satisfying self-interests (Mayes & Allen 1977, 675).[5] All these definitions describe the negative view of politics as unpleasant, conflict-based, malicious, and self-serving. Pinto (1994, 111–130) highlighted the following behaviors that illustrate negative politics:[6]

- Behavior that focuses on gaining benefits for the individual or a group at the organization's expense

- Behavior that circumvents the organization's legitimate power structure

- The use of means not endorsed by the organization to achieve approved ends, or the use of approved means to obtain unapproved goals

3.1.1. Attributes of Negative Politics

> *You live with your thoughts . . .*
> *so be careful what they are.*
> —Eva Arrington

Negative politics are based on self-centeredness and working toward satisfying self-interests. People who like negative politics do not trust others and want to win at any cost, often at the cost of others. They take advantage of others' weaknesses and want to control people and resources to meet their objectives. Table 3.1 lists the attributes of negative politics and their purposes.

Finger-pointing or blaming others

Pointing fingers, or blaming others, when something goes wrong leads to negative politics because people feel defensive and blame back. Pointing fingers is nonconstructive because problems and issues are not openly addressed. Neither party is prepared to accept accountability and responsibility for mistakes. People find scapegoats to defend themselves

Attributes	Purposes
Finger-Pointing or Blaming Others	Deflecting accountability
Control	Controlling people, budget, and other resources
Hidden and Competing Agendas	Pushing personal benefits
Subjective Decisions	Making decisions without thorough analysis
Favoritism	Increasing expectations for reciprocity
Functional Silos	Avoiding collaboration
Retribution/Revenge	Taking revenge for past events

Table 3.1. Attributes of Negative Politics and Their Purposes

This behavior passes the blame to others, deflects accountability, and leads to poor relationships and negative politics.

Control

Control is a common attribute leading to negative politics. *Control* means managers want to control people, budget, and other resources. It leads to negative politics because most people resent being controlled and being deprived of information and resources they consider useful to carry out their project activities.

People are willing to make a sincere contribution and are motivated to increase their performance if they think their managers believe in them, and they are trustworthy. They feel discouraged when they discover their managers are interested in using their formal power and controlling them, their resources, and their activities. Therefore, they do not work as a team and lose interest in achieving project objectives. The overall performance of the project team and project personnel goes down, leading to risks in completing projects on time and on budget. Psychologically, the trust level goes down, and miscommunications and conflicts happen, which create more problems and challenges.

Hidden and competing agendas

Project teams consist of members with diverse backgrounds, expectations, norms, cultures, and interests. People possibly have different approaches for doing things to meet project objectives. Competing agendas can even be healthy for a project's progress as long as there is open communication and willingness to share, discuss, and evaluate pros and cons of different viewpoints.

Negative politics arise when people do not openly discuss their ideas, their processes, and their techniques for managing projects. Often, people go to any extent to push their agendas and play politics to gain the support of many project personnel to meet their objectives, even at the cost of organizational objectives.

Disturbingly, people who use their powers to do their favorite projects use resources to meet self-interests and push personal benefits rather than furthering organizational goals and strategies. Organizations greatly increase their performance if unity of purpose is established, and people work together to meet common objectives. Negative politics are caused, not necessarily by competing agendas, but especially when those agendas are hidden and not shared openly with other project personnel.

Subjective decisions

Most people get upset when decisions affecting them are made without thorough analysis. It makes them suspect decision-makers' motives, and it can create unnecessary misunderstanding. Such decisions also might be bad for the project or organization. Stakeholders might not support such decisions and might negatively influence outcomes. Senior management members might use this behavior to push their agendas or satisfy their self-esteem. People are also unhappy when management avoids promptly making decisions that might affect project progress, or decisions are made without proper input from stakeholders.

However, if such decisions must be made because of time constraints and other valid reasons, decision-makers must clarify the situation and their exact reasoning behind the decisions made. Lack of open communication and decision-making without thorough analysis reduce trust and confidence in the decision-makers' ability and leads to negative politics.

Favoritism

Favoritism is often seen as an unfair and poor management practice that increases expectations for reciprocity. Managers might use favoritism to assign high-profile projects, give roles and responsibilities that speed individuals' progress, and distribute rewards. People often get upset when rewards are based on favoritism and subjective criteria. They talk about such decisions negatively among their groups and team members and strongly condemn such decisions and behaviors. Such behaviors are also perceived as a lack of good management skills and practices.

Repeated acts of favoritism by management reduce project team morale and set a bad management example. Therefore, rewards and recognitions must be based on objective criteria to avoid creating poor morale and negative politics.

Functional silos

Functional silos lead to negative politics. Many people are too interested in pursuing their ideas and work in isolation to avoid collaboration. They do not share and discuss their ideas with others and, sometimes, do project activities that do not align with organizational strategies and goals. Many functional silos are subject matter experts who could contribute much by sharing their ideas with others or learning from one another. They might display this behavior because of insecure feelings and fear of rejection. They spend their time and resources in isolation and use the resources, information, and processes inappropriately. Moreover, the organization does not gain from their expertise, special experiences, and knowledge.

Retribution or revenge

Feelings of revenge represent the dark side of politics, which is negative for the project and organization. Many people in organizations carry the baggage of negative experiences and memories. Some project managers have experienced working with people from whom they had less cooperation and understanding. Those people possibly did not put in their best effort for their project managers for reasons not under their control.

Project managers might remember those poor performers, and if those people are working on their team, get even with them and get involved in negative politics. These project managers believe those people did not cooperate with them and did not put their best effort intentionally on past projects, so these project managers take revenge on those project personnel, making situations even worse.

The same scenario can also be viewed from the team members' perspective. Team members might be unhappy with the leadership and management skills of project managers with whom they worked on previous projects. If they work with the same project manager for a new project, they remember the unpleasant experiences they had with that project manager and develop feelings of retribution and revenge and do not contribute to team efforts and work as hard as they could. Their commitment level and motivational level are so low that they even go to the extent of sabotaging other project team members' efforts, leading to many problems and increased risk of project failure.

Such feelings of revenge and retribution can be easily resolved if people show their maturity level by coming forward and apologizing sincerely for past mistakes and unpleasant behavior to start things with a clean slate. Such maturity can be attained from senior management's proper support, mentoring, and leadership.

Politics can be both positive and negative depending on how they are used. Negative politics focus on self-centeredness and gaining control and power over people and resources. People who use negative politics have their agendas and a compelling desire to win all the time at all costs. Negative politics have the following attributes: finger-pointing (blaming others), control (over people and resources), hidden and competing agendas (emphasizing self-interests), subjective decisions (lack of competence), favoritism (assigning roles and distributing rewards), functional silos (working alone and not sharing information with others), and feelings of retribution or revenge. Senior management should create an environment to discourage negative politics in organizations.

3.2. Positive Politics

There is a positive side to everything. It just takes a positive mind to see it.
—Anonymous

Many people believe politics are negative because they lead to resentment, anger, confusion, and frustration. They believe most people use politics to meet self-interests and objectives. Therefore, most people do not want to engage in politics and even urge their team members to stay away from politics and the people who play them.

Positive politics emphasize the dynamics of politics and use effective management and leadership techniques to open communication, challenge processes, gain trust and commitment, and share information. Extraordinary results are achieved through collaboration, cooperation, and synergy.

Politics can be positive, and if managed well, they can help project managers meet project goals and organizational goals effectively. Because politics cannot be eliminated, it can be argued that senior management should even create positive politics because they will redirect the energy of project personnel from the negative side to a culture of more commitment and teamwork.

3.2.1. Ten Commandments of positive politics

Positive politics rechannel people's energy and creates a culture of better teamwork, leadership, and communication (TLC).
—Vijay Verma

Often, politics have a negative connotation and the concept of positive politics is less understood. Politics are part of life in project environments, so project managers cannot ignore or avoid politics. Instead, they should accept that politics are simply a way to get things done through others in any organization. Therefore, they should look at politics positively and use positive politics to gain support and cooperation from their project stakeholders.

We must understand what positive politics are and what factors and key ideas lead to positive politics. This section addresses these issues to help project managers understand the positive aspects of politics and apply these to deliver successful projects. Table 3.2 shows the Ten Commandments of positive politics and their main purposes.

	Commandments	Purposes
1	Develop Organizational Strategies and Goals	To establish, communicate, and gain buy-in
2	Build Honest Alliances	To increase cooperation
3	Focus on Collaboration, not Competition	To avoid negative politics and win/lose emphasis
4	Develop Synergy and Teamwork	To increase overall performance
5	Emphasize High Quality	To invigorate people to do the right thing the first time
6	Provide Appropriate Support and Training	To help people grow their potential
7	Encourage Sharing of Information	To increase synergy
8	Design an Objective Reward System	To reduce favoritism and subjectivity
9	Develop an Effective Mentoring System	To increase overall knowledge and confidence
10	Embrace Cultural Diversity	To accept and capitalize on cultural differences and dynamics of multi-generational workforce

Table 3.2. Ten Commandments of Positive Politics and Their Purposes

1. Develop organizational strategies and goals.

Developing organizational strategies and goals is the most important idea of positive politics. Many project managers recognize that often, there is much confusion, lack of clarity, and misunderstanding about organizational strategies and goals. They try to find information from their sources to understand and confirm organizational goals and objectives. They use a combined formal and informal network to find relevant information about organizational strategies and plans.

Lacking correct information directly from senior management, they might assume organizational goals. Such assumptions might be wrong, and they then try to meet those goals, leading to the wrong use of resources. Here, senior management can help lift the fog and communicate clearly the goals and objectives of the organization and relevant programs and projects. Senior management should use a three-step process to establish goals, communicate goals, and gain support of stakeholders to meet those goals and create unity of purpose:

- **Establish organizational strategies and goals.**

  Senior management must follow this first step to create synergy in the organization. They must establish long-term strategies for the organization and then define relevant programs and projects to meet those strategies. They should keep in mind the associated business, financial, and human constraints.

  This step might be most effective if members of senior management, such as the president, executive vice presidents, and vice presidents, are consulted. Conflicts and misunderstandings must be addressed and resolved to ensure organizational strategies and goals align with business goals and that they can be accomplished with available work force and financial and other resources and constraints.

- **Communicate strategies and goals.**

After strategies and goals are established, the senior management team must communicate them clearly to the rest of the organization to create unity of purpose and common direction. This effort is a major challenge because lack of clarity in communicating goals and objectives leads to confusion, conflicts, and misuse of resources. Management should use combined effective communication styles, as needed. They can use all-hands meetings to convey the message, or each department head might communicate the organizational goals and objectives to people in their department.

Each department head must consistently use clearly written strategies and goals. People should be encouraged to ask questions and raise their concerns, if needed, about the goals and objectives for clarity and realism. Management must address all questions and comments clearly and professionally rather than be defensive about the issues. A lack of open communication at this stage creates confusion, misunderstanding, and a lack of synergy.

- **Gain buy-in from stakeholders.**

After strategies and goals are communicated to stakeholders, you must gain their buy-in and commitment so they do everything possible to meet the strategies and goals. Senior management should gain commitment from primarily a minimum of two levels of management, which might include executive vice presidents and vice presidents. Then, the vice president should take responsibility for communicating, clarifying, and addressing comments to gain commitment from people in the next level of organizational hierarchy, which might include directors, functional managers, resource managers, program managers, and project managers.

Gaining true buy-in is difficult, but once it is achieved, the dividends are high. Therefore, senior management must invest sufficient time and energy to gain commitments at various levels.

The steps to gain commitment can be described as follows:

- **Genuine involvement**

  Genuine involvement is the first and most important step to gain commitment from people in the organization. Management must recognize that involvement must be genuine, which means people must feel comfortable in raising their viewpoints, even if they differ from those of senior management. They must be assured they are listened to, and their comments will be appropriately addressed. They should expect timely feedback in an open and constructive manner.

  Often, managers involve people, but it is not perceived as genuine. Senior management should be serious about involving people genuinely rather than just giving lip service about involvement. If people believe management is insincere in seeking participation from them, they do not participate wholeheartedly, and management loses an opportunity to gain their viewpoints and, hence, their cooperation and commitment.

Sometimes, if people are not involved genuinely, things get even worse as they feel unhappy and then resent management's superficial behavior. These feelings lead to lack of commitment, synergy, and feedback associated with organizational strategies and plans to meet those strategies.

- **Acceptance and agreement**

  Gaining acceptance is the next important step in gaining commitment from stakeholders. Management can only hope to gain acceptance if they are perceived as genuine in getting feedback and comments from people in the organization.

  Stakeholders should be given enough time to think about organizational strategies and goals and, especially, management perspective and priorities. Then, the quality of feedback and comments are well thought out, and often, stakeholders think of the challenges and even the solutions to meet those challenges. This approach not only encourages active stakeholder participation, but also gains their acceptance of organizational goals and objectives. It is nice, but not essential, to get consensus from all stakeholders, but it is crucial to reach the agreements they are committed to support.

- **Commitment and buy-in**

  Once stakeholders agree with management about the strategies and organizational goals, they are more likely to support and cooperate wholeheartedly to help meet those goals and strategies. Acceptance and agreement are reached after open communication and evaluation of total feedback and concerns from stakeholders. To gain commitment, project managers should ensure the agreements reached are carried through, and to increase the probability that the agreements are carried through, the following three conditions must be met:

  1. **Agreements must be free of fear, threat, or coercion.**

     Agreements made under fear or threat have low probability of being carried through. The party feeling the fear or threat will do everything possible to get out of the agreements reached.

  2. **Agreements must be fair and reasonable.**

     There should be complete openness, and neither party should try to manipulate or trick the other party into accepting unfair and unreasonable agreements. In

111

such cases, even if the agreements are formally reached, one party might think he or she has been manipulated, and will find means to get out of the agreements.

3. **There must be a meeting of the minds.**

This condition is important and subtle. This "meeting of the minds" means both parties have good mutual understanding and consider their relationship more important than just achieving their goals. Communication and conflict management are easy when two parties have good understanding. Both parties try to be considerate to each other's interests. Negotiations are done in good faith to reach win-win solutions.

It can be argued that the third condition incorporates the main elements of the first two conditions. If there is good understanding between both parties, then they are like-minded in ethics and values, and they will not use fear, threat, or coercion to force the other party to accept unfair and unreasonable agreements. Both parties treat each other the way they want to be treated themselves.

If the three conditions are met while communicating strategies, stakeholders are genuinely involved, and their comments and feedback are appropriately addressed, stakeholders will commit to cooperate with management. Furthermore, senior management is responsible to give proper rewards and recognition to stakeholders to motivate and reinforce them positively long term to meet organizational strategies and goals. These actions avoid misunderstandings and conflicts and discourage negative politics. Figure 3.1 shows a model to gain commitment from and motivate stakeholders to deliver successful projects.

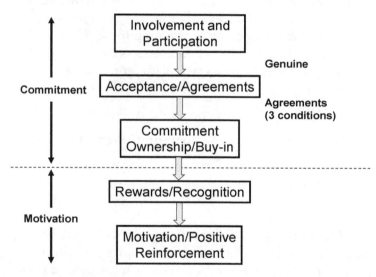

Figure 3.1. Commitment and motivation model.

2. Build honest alliances.

Often, negative politics happen because people do not trust one another. Building honest alliances helps create sincere, long-term business relationships and increase cooperation. In project management, it is particularly important because stakeholders might have different interests and expectations. Building alliances is more challenging when stakeholders do not clearly define their requirements and expectations.

In such circumstances, project managers must not only meet the specified requirements identified in the project's scope, but they should also uncover their clients' unidentified needs and work together to address those unidentified needs. This effort requires good communication skills and the ability to ask probing questions that trigger the client to think "out of the box" and identify more requirements useful for them to meet in the long term. Uncovering and addressing unidentified needs is an art, which can be achieved by demonstrating genuine concern for the following:

1. The client's advancement and growth

2. Future business opportunities for the client

3. Long-term performance improvements for the client

The main theme of the three points is to give highest priority to looking after clients and their interests. Negative politics arising from lack of trust and a tendency to take advantage of the other party decrease. Sincere attempts by project managers to build honest alliances with their stakeholders and clients lead to more trust and cooperation and, hence, to positive politics.

3. Focus on collaboration, and not on competition.

Creating a climate of teamwork and collaboration is challenging because project teams consist of people with diverse interests and expectations. In project management environments, all team members and other stakeholders must work as partners. Management should create a climate of focusing on *collaboration* rather than on *competition* to avoid negative politics and win-lose emphasis. Sometimes, competition can be healthy, but in many situations, it leads to negative politics because everyone wants to win. Competition focuses on winning, even at the cost of others. Collaboration is more important and powerful in achieving cooperation and trust than a strong desire to win.

The concept of partnership should be used more often in project management. Team members should be called partners, which implies that all team members gain or lose together. For example, in a simple business relationship, if two parties are 50-50 partners, they share profits and losses equally. One partner cannot think of passing 100 percent of the loss to the other partner because they are equal partners and must share the negative and positive business outcomes equally.

The concept of competition among different project managers suggests applying strategies to win and trying to achieve your project objectives even at the cost of other higher-priority projects. A common example is to use politics to get organizational resources for your project's benefit rather than for the organization's benefit.

115

Management is primarily interested in meeting overall organizational goals and objectives, and not just those of a few projects. Therefore, management should focus more on collaboration than on competition to optimize organizational resources to meet overall strategies and goals. The competition approach creates a desire among project personnel to win even at the cost of others, which is a win-lose approach and leads to negative politics. A collaboration approach emphasizes helping one another win and encourages people to think win-win solutions, leading to positive politics and increasing overall organizational performance.

4. Develop synergy and teamwork.

Teamwork is critical to project success. Most teams are composed of members with diverse backgrounds, expertise, beliefs, and expectations. Sometimes, a few team members might not be active team players and even become barriers to teamwork, especially if they don't like their project manager, team leader, and the work environment. In such circumstances, they play negative politics, which reduces collaboration and synergy. They are not motivated to do their best and do not cooperate with other team members to meet common objectives. In some cases, they withhold information, do not share knowledge and information with other team members, and even intentionally become obstacles to overall team success.

Senior management can be proactive in promoting the importance of teamwork and synergy. They should urge program managers and project managers to build effective and high-performance teams.

Project managers must involve their team members in all major decisions to gain their acceptance and commitment. Senior management should provide proper resources and team rewards for producing high performance consistently. They should encourage project managers to develop and implement practical guidelines for effective team building (Verma 1997, 133–148).[7]

This approach not only discourages negative politics, but also leads to a culture of increased collaboration and teamwork and increases overall performance. Team members help one another produce their best and resolve conflicts and problems as they arise. They have more confidence in senior management and feel inspired to work together to meet organizational goals and objectives. Such work environments that focus on synergy and teamwork minimize the negative impacts of politics.

5. Emphasize high quality.

Quality is crucial to project success, and commitment to high-quality standards should be incorporated in the organization's vision and mission statements. A lack of quality can destroy the reputation of the project manager and the organization. Therefore, project managers and team members must recognize the importance of delivering projects with high-quality standards and meeting scope, time, and budget constraints.

Project managers and team members must anticipate, address, and resolve all internal and external issues that might lead to compromises in quality. Often, when the quality of deliverables is poor or unacceptable to a client, people point fingers at one another or blame others for problems that could have been easily avoided and resolved if people worked together and cooperated with a commitment to produce high-quality outcomes. In a sense, this redirects the efforts of project personnel from negative politics to a climate of more collaboration and cooperation.

Three factors help improve the quality of project deliverables:

1. **Pride of ownership**

   This factor is most important in meeting high-quality standards. Team members and all project stakeholders should take ownership of their jobs. They should feel pride in fulfilling their project activities with high quality. Usually, if this pride of ownership is missing, people try to do tasks without paying attention to details and overall quality implications. In such cases, if anything goes wrong, people point fingers at others and deflect issues that led to lower quality.

   Senior management must create a culture of self-accountability, trust, and self-motivation. Project managers must help their team members expand their potential, showing trust and confidence in them so they do not hesitate to accept new challenges. This pride of ownership is difficult to instill in team members, but the dividends are high for improving the quality of project deliverables. Table 3.3 describes PRIDE (in a nutshell).

	Keyword	Focuses On
P	Professionalism	Communicating with a positive attitude and striving to understand people and their requirements
R	Responsiveness	Responding to requirements promptly
I	Innovation	Creativity and thinking outside the box
D	Dedication	Fully committing to produce high-quality outcomes
E	Efficiency	Optimizing use of available resources

Table 3.3. PRIDE (in a Nutshell)

2. Commitment as opposed to compliance

Project managers must recognize the difference between commitment and compliance. Meeting compliance requirements only leads to bare minimum performance to meet specified scope and requirements. Sometimes, project personnel are not motivated enough to give their best if they are unhappy with the work environment and the job.

It also depends on how project managers use their power (formal or informal) to get things done. If project managers use their powers formally, project personnel are more likely to resist, and at most, just meet specified scope and produce only bare minimum, that is, specifically what is asked to be done.

Quality will be low if project personnel are only interested in meeting compliance requirements rather than committing to high performance. When project personnel are committed, they do their best and go beyond specified scope and requirements. They put in extra effort to add more creativity and innovation that lead to higher quality. Often, committed people are also self-motivated. The highest level of commitment also leads to more flexibility, better work attitude, and increased teamwork to produce high-quality outcomes.

3. **Prevention as opposed to detection (Doing things Right the First Time (DRFT))**

High quality means Doing things Right the First Time (DRFT). This concept focuses more on prevention than on detection. It implies that project personnel should anticipate and resolve problems so they can prevent problems and issues that lead to poor quality outcomes.

Unfortunately, many manufacturing and production processes do not emphasize prevention, but deal with the quality problems after the poor quality products are finished. This is an ineffective approach because the damage in wasted effort and money has already happened, leading to increased overall product cost. There are only two options left at this stage:

- Redo the whole thing over

- Repair and rework to resolve quality problems

Preventing quality problems before they occur requires working together with high commitment from everyone involved. In an engineering and manufacturing environment, this can be achieved by using the concepts of Concurrent Engineering (CE) where engineers, manufacturing personnel (machinists and so on), and assemblers work together and review the whole process from design to finished product before detailed engineering and production happen. Problems or risks in producing high-quality products are anticipated and resolved before significant effort and money are spent.

With this approach, the overall costs of making changes to improve product quality are minimized because major changes are made at the design stage when the cost of making changes is low and benefits are high.

Senior management should focus on high quality, which should be part of the mission statement. They should urge project managers and team members to use three techniques outlined in this section:

1. Pride of ownership

2. Gaining commitment rather than just compliance

3. Doing things Right the First Time (DRFT) to produce high-quality products.

These actions reduce finger-pointing, which leads to negative politics. Instead, senior management should create a sense of ownership, leading to personal satisfaction and positive reinforcement, which contributes to personal growth and that of the organization.

6. Provide appropriate support and training.

Senior management must provide support for all their employees at different levels to help them do their jobs more effectively. This prevents people from getting discouraged and failing to produce their best performance. The required support can be provided at various levels:

- **Professional level** — At this level, management must commit to provide appropriate training, as needed, to increase their employees' professional performance, including training to carry out the job's technical parts and to develop and implement project processes. Management should provide appropriate training to project personnel at various levels to increase the following skills as they progress from entry-level positions to project managers, program managers, and portfolio managers:

- **Technical skills** — Management should provide technical training so employees can do the technical part of project processes more efficiently. Sometimes, it includes basic training about general business activities and processes to carry out project activities (project management processes, templates, software packages, and so on).

- **Interpersonal skills** — After people feel comfortable with the technical part of their job, they are ready to take on supervisory roles. Besides technical training, management should provide training to enhance their interpersonal skills to increase their ability to communicate, motivate, negotiate, make decisions, influence, and manage conflicts. This training in soft skills helps them manage their people and projects more effectively.

- **Leadership skills** — As people progress from managing projects to programs and departments, they need more leadership abilities besides their interpersonal skills and technical skills. They must be able to look at the big picture and analyze project outcomes in business results. To advance in most organizations, project managers must recognize the following:

  - People are *hired* based on their technical skills.

  - People are *promoted* based on their interpersonal skills.

  - People are *fired* because of the lack of leadership skills.

- **Organizational level** — Senior management commits to provide support to their program and project managers at an organizational level. This increases their confidence in senior management and convinces them that management is committed to help them meet organizational goals and objectives. Support at the organizational level might include the following:

- Providing sufficient resources with proper skill mix

- Providing resources at the appropriate time to meet deadlines

- Providing support from various internal service groups

- Providing support from external stake-holders (clients, regulators, and so on)

- Providing proper financial resources, as needed, to meet budget constraints

- Providing support to mitigate project risks (unexpected delays and cost overruns)

- **Personal level** — *Personal level* refers to providing support at a personal and an intrinsic level. Sometimes, project personnel have personal problems related to health, family, and work-life balance. They might have issues with ethics, social responsibility, and their organization's value system. This can become an obstacle for people in exploiting their potential optimally.

  Senior management must recognize such problems and address them openly and promptly. Management should realize issues might be personal and sensitive, and therefore, they should try to see things from their project managers' perspective. Project managers should also do everything possible to help team members resolve their personal issues to maintain high performance.

7. Encourage sharing of information.

Information is an important factor leading to negative or positive politics. Some people believe withholding certain information and not sharing it with others readily increases their power. People withhold information to show their power, which leads to negative politics. Some people do not share information, even if other team members need that information to carry out project activities assigned to them. Many project personnel resent this behavior and feel frustrated because of the lack of information that could be available to them if those team members did not withhold it. Instead, if the information is shared with the team, it increases overall knowledge and expertise.

Management should encourage sharing of information, and people should be rewarded for sharing it to create positive politics. In addition, management should discourage habits of withholding information and, if necessary, should even punish those who withhold information to increase their power.

Withholding information can be viewed as a lose/lose situation. For example, if John withholds information from other members on the team, they are also unlikely to share their information with John, and John's information power goes down. Project managers must recognize that information is power that "the more you share, the more you get." Therefore, senior management must create a culture that encourages "sharing of information" rather than "withholding information." This culture promotes more collaboration, helping one another become more knowledgeable and reducing negative politics and frustration among team members.

8. Design an objective reward system.

Reward systems, if designed and implemented incorrectly, can lead to negative politics, resentment, anxiety, and frustration. Senior management must understand the definition of a reward and recognize the dynamics of reward systems and their perception individually by people working on a project or as a team member. In addition, management must design a reward system that is objective and based on people's performance rather than using subjective criteria.

A reward is considered effective if 1) it satisfies the **need** of the recipient of a reward, and 2) that need is one the recipient **values most**. For example, management gave a monetary reward to team members because they thought it satisfied the members' need without understanding whether that need was important to them. In this case, that reward would not be perceived as effective if it did not satisfy the need they value the most.

The following are questions you should answer when establishing an effective reward system:

1. How do you learn what needs people value most?

   Different people have different needs, and needs might change with time, even for the same person.

2. What rewards would satisfy the needs they value the most (matching rewards to needs)?

After management understands these two questions, they should develop a reward system based on objective criteria. It is better if they involve managers from different departments and seek their input in designing the reward systems. Most people might not mind if they do not get the rewards, but they would resent the reward system when certain team members get the rewards based on favoritism and other subjective criteria. This action results in a lack of trust and respect between management and employees. It also reduces the level of self-motivation and cooperation among team members and increases misunderstandings, conflicts, and frustrations, leading to negative politics.

9. Develop an effective mentoring program.

Mentoring is an excellent technique to neutralize negative politics and create a culture of collaboration and cooperation. It is even more important in the context of politics because the dynamics of politics are complex.

Managing politics effectively requires many years of practical experience working with different people in different organizations. Management can provide this learning experience by developing an effective mentoring program in which experienced managers are assigned as mentors to new and upcoming project managers and team leads. The mentors are responsible to teach the tricks of the trade to new project managers and help them solve problems as they arise.

The mentoring program can be formal or informal. The human resources department runs a formal program where the roles and responsibilities of both the mentor and the mentored are defined, and regular meetings are held between them. The human resources department might encourage both parties to develop action items to be learned and applied and then follow up regularly. Any department can run an informal mentoring program, and mentors are voluntary. New project managers are encouraged to meet with experienced people willing and motivated to share their experiences and knowledge with others. Mentors are committed to help new project managers by giving them guidance, advice, and honest feedback.

A mentoring program is an efficient and excellent way to learn new practical techniques and skills in a nonthreatening environment. In many cases, the mentors become champions and allies for many project managers they mentored. The young managers feel comfortable in discussing both their strengths and their weaknesses with their mentors because their mentors are genuinely interested in helping them succeed and progress in the organization. This program also increases the trust and respect between the mentors and mentored. The program can be reinforced positively if management provides enough organizational resources and gives proper recognition to mentors for their efforts and willingness to share their expertise.

10. Embrace cultural diversity.

Projects are done by project teams consisting of people with diverse beliefs and cultural backgrounds. Cultural backgrounds are important in defining the dynamics of people working together on a common project. *Culture* refers to a commonly shared set of values, beliefs, knowledge, and attitudes. According to Hofstede (1993), culture can be described as *mental software*— the collective programming of the mind that distinguishes members of one group of people from another.[8] Global projects often involve joint ventures and team members from various countries. Project managers must know major elements of culture and people's responses to cultural differences. They must appreciate that cultural differences are a reality of today's global economy, and therefore, they must learn to capitalize on cultural differences.

Culture is difficult to define and complex because of its several dimensions, variables, and elements. Martin (1993) identified seven major elements of culture: [9, 10]

1. **Material culture** — Includes physical objects and technologies created by people and tools, skills, work habits, and attitudes toward work and time.

2. **Language** — Represents a mirror of culture and a primary means of communication. Includes gestures and expressions that might be interpreted differently by others, even within the same language.

3. **Aesthetics** — Refers to art, music, literature, and related customs and artifacts.

4. **Education** — Refers to how knowledge is transmitted and how people approach problems and relate to others.

5. **Religions, beliefs, and attitudes** — Represents an important complement of culture. Religion is the mainspring of culture.

Beliefs and attitudes influence other elements of culture, including dress, eating habits, attitudes toward work, and punctuality.

6. **Social aspects** — Includes forming coalitions, unions, and groups to accomplish common goals and classes of society, social clubs, family relationships, and other groups that influence attitudes and values about life.

7. **Political life** — Becomes important when projects are global, and many governments are involved. Represents the concerns of various governments about the profits, legality of transactions, number of jobs created, treatment of people, safety, and environmental concerns.

Hofstede (1993, 14) identified major critical dimensions of cultural differences:[11, 12, 13]

- **Power distance** — Refers to how people deal with inequality and differences in organizational hierarchy.

- **Gender** — Refers to societies dominated by masculine and feminine genders.

- **Individualism-collectivism** — Refers to how people behave individually or in groups.

- **Time** — Refers to time horizon, the value of time, and focus.

- **Attitude toward life** — Refers to a value system based on quality of life, as opposed to the desire for material needs.

- **Uncertainty avoidance** — Refers to how people respond to uncertainties and ambiguities in the workplace.

Other elements of workplace diversity include people in a minority or physically challenged in different ways. In some countries, laws relate to how physically challenged people must be accommodated in organizations. Such people feel emotionally hurt when they are not treated equally with others in organizations, so they might show a lack of cooperation while working in teams.

In some countries and societies, there are special privileges for minorities and physically challenged people. Sometimes, project managers discriminate when assigning project tasks to various team members with physical challenges. Senior management can prevent such problems by creating better awareness of workplace cultural diversity and urging their program and project managers to treat everyone equally and capitalize on cultural diversity rather than getting hung up on differences in culture.

Management should provide enough training to their project managers about managing in a culturally diverse work environment. Another dimension of cultural difference is the multigenerational workforce. There are significant differences among the organizational value system, job security, work attitudes and ethics, work-life balance, and so on, between the use of technology and social media, baby boomers, and generation X, Y, and millennials. Senior management must provide proper training for programs and project managers to manage project teams with diverse multigenerational team members, leading to more collaboration, synergy, harmony, and mutual respect.

Although people usually think negatively of politics, they can be positive. In a positive sense, politics can be viewed as a natural way to get things done in organizations by dealing with one another effectively. Positive politics emphasize "We" and the importance of working as a team. The focus of positive politics is to find ways to create synergy and gain support and cooperation from project stakeholders, which can be accomplished by following the Ten Commandments of positive politics.

Chapter 3 Summary

People generally view politics as negative, but they can also be positive. In a positive sense, politics are a natural way to get things done in organizations by dealing with each other effectively. Positive politics emphasize "we" and the importance of teamwork. Positive politics focus on finding ways to create synergy and gain support and cooperation from project stakeholders.

Negative politics focus on satisfying self-interests whereas positive politics focus on working together for the benefit of the organization. Negative politics have the following attributes:

- Finger-pointing or blaming others to deflect accountability

- Control over people, budget, and other resources

- Hidden and competing agendas to push personal benefits

- Subjective decisions without thorough analysis

- Favoritism to increase expectations for reciprocity

- Functional silos to avoid collaboration
- Retribution/revenge for past events

In contrast, you can follow Ten Commandments to make politics positive:

1. Develop organizational strategies and goals to establish, communicate, and gain buy-in.
2. Build honest alliances to increase cooperation.
3. Focus on collaboration, and not competition, to avoid negative politics and win-lose emphasis.
4. Develop synergy and teamwork to increase overall performance.
5. Emphasize high quality to invigorate people to do the right thing the first time.
6. Provide appropriate support and training to help people grow potential.
7. Encourage sharing of information to increase synergy.
8. Design an objective reward system to reduce favoritism and subjectivity.
9. Develop an effective mentoring system to increase overall knowledge and confidence.
10. Embrace cultural diversity to accept and capitalize on cultural differences.

Negative politics thrive on command and control, whereas positive politics encourage effective communications, influencing, and teamwork. As a first step toward positive politics, senior management should establish clear organizational strategies, communicate and seek genuine feedback from program and project managers, and address their concerns to gain support and ownership. This process provides a clear sense of purpose and direction. Senior management should try to create positive politics by using the Ten Commandments, which create a work environment to increase synergy, collaboration, and cooperation.

Why Organizational Politics?

Skillful politicking and power brokering are
critical to successful project management.

—Vijay Verma

Every organization has politics that vary in nature and intensity. Some politics are associated with project management discipline and people's behavioral issues, whereas others are general because of organizational issues, management issues, and global economic factors.

The major issues that lead to organizational politics relate to the following four categories:

1. Project and Project Management Issues

2. People Behavioral Issues

3. Organizational and Management Issues

4. External and Global Market Issues

These issues are shown in figure 4.1.

135

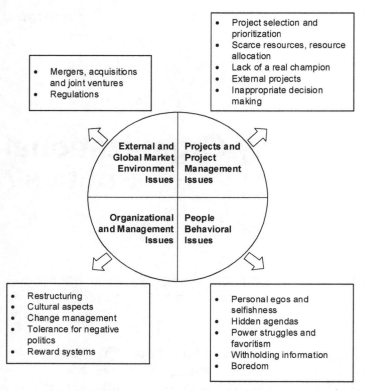

Figure 4.1. Four main issues leading to politics.

4.1. Projects and Project Management Issues

Before you can begin to think about politics at all, you have to abandon the notion that there is a war between good men and bad men.
—Walter Lippmann

Politics affect the outcomes for many projects. Project managers must know those politics so they can deal with them to minimize negative impacts. The following are common politics related to projects and the project management environment:

4.1.1. Project selection and prioritization

Project selection and prioritization often lead to politics, especially negative politics, because every organization has limited capacity in financial and human resources. After developing organizational strategies and goals, senior management must select a portfolio of programs and projects that align with those strategies, meaning that many projects might have to be rejected because of limited organizational capacity. They must select a few relevant projects and develop a prioritization system so the right programs and projects are done at the right time.

Because market and economic circumstances change, senior management must review the portfolio of programs and projects continually and decide their priorities to meet overall organizational goals. They must be willing to change strategies, if needed, and terminate projects that no longer align with organizational strategies under new market conditions.

Management must develop clear guidelines and a process for selecting projects and deciding on the priorities of selected programs and projects. They must base this process on well-thought objective criteria. Lack of a clear process and a subjective prioritization system often leads to negative politics.

Many project managers feel frustrated when there is no such established process or system, and resources are allocated subjectively to favorite projects. In such cases, people use politics to push their favorite projects at the cost of those that should be given higher priority to meet organizational strategies and goals; so, the wrong programs and projects consume organizational resources. Such organizations change strategies and direction for the wrong reasons, and eventually, they might not survive in the marketplace.

4.1.2. Scarce resources

Resources are limited. Many project managers believe they do or do not get the proper resources because of politics. They use their influencing skills, informal power, and politics to get resources they need to meet deadlines for their projects because they often compete for the same resources. If an organization has unlimited resources, then all project managers can get the resources they want, and there are no politics over resources. But the truth is that resources are limited, and everyone fights for them in their way, which leads to politics and conflicts among departments and managers. Everyone uses politics to get what they want, even at the cost of the organization.

4.1.3. Resource allocation process

Many project managers believe the decision process for resource allocation is political because it is highly subjective. There are usually no rules and procedures to make decisions, and if there are any, they are too complex and often unwritten. If they are written, they change.

Project managers feel frustrated in this regard because they don't know the rules, procedures, and decision-making process that management uses to allocate resources. Even if and when they think they know the decision-making process, the decisions change without proper analysis and evaluation. The decisions are also highly subjective and not based on widely known objective criteria, so project managers use their political skills to influence the decision-makers to get the resources they want for their project, even if it does not benefit the organization.

4.1.4. External projects

Many project managers believe they can establish good relationships, formal and informal communication, and good understanding with stakeholders for internal projects. However, for external projects, they don't think they know the external organization's political landscape and dynamics. Therefore, they are unaware of the external organization's political structure, informal relationships, and politics that can easily affect their project's outcomes.

Client representatives for external projects are sometimes either noninfluential or lack the senior management support needed to resolve problems and issues promptly. Therefore, project managers face several difficulties in managing external projects and feel frustrated because of lack of support from the external organization.

4.1.5. Lack of "real" champion

Lack of a "real" champion is perhaps the most important cause of project problems and politics for internal and external projects. Before understanding this issue, we must remember that a champion for the project and the project manager must be the following:

- **Person with a vested interest** — Champions must have a stake in the project and want the project and the project manager to succeed because they also benefit from a successful project outcome.

- **Remover of roadblocks** — Effective project champions should be able to remove project obstacles by using an appropriate mix of organizational authority and influencing skills, which are important in dealing with politics. Champions must have organizational clout and influencing skills, including the appropriate connections and relationships (network power) to remove roadblocks encountered in the project.

  Champions either should have sufficient power (formal and informal) themselves or should have easy access to another senior person with the power to help remove obstacles. Champions must have the right combined interpersonal skills and ability to evaluate when and how to use their power effectively. To help with politics, they must be willing to "go to bat" for the project managers or advocate their concerns and help make things happen in their favor.

Champions might or might not be directly involved in the project daily, but they must perform both these roles effectively. Project managers often raise the following five questions about champions and their roles, especially about removing obstacles:

1. Who are the real champions?

   Because the first role relates to vested interest, many people think project sponsors are champions because they provide funding for the project. However, project managers must be cautious that project sponsors might not be effective champions, especially when they are expected to fulfill the second role of remover of obstacles. Project sponsors might be ineffective for the following reasons:

   • Lack of proper organizational authority or positional power

   • Lack of appropriate mix of interpersonal skills, influencing skills, and network power

   Project managers should be aware that their sponsor might not be a strong champion, and therefore, they might have to develop a strategy to resolve the problem. Project managers recognize this scenario when they run into a difficult political situation beyond their expertise and authority, and their sponsor cannot resolve the issue successfully because of the aforementioned reasons.

2. Where are the real champions?

   This question refers to an appropriate level in the organizational hierarchy. Project managers should find potential

champions with appropriate positional power
because champions should have sufficient
organizational formal authority (legitimate
power) to remove obstacles. Project managers
must recognize that despite their good
technical project management knowledge and
expertise, they cannot help the champions
directly with positional power issues
because either the champions have sufficient
legitimate power, or they don't.

The best strategy for project managers in
this regard is to find champions already at
the appropriate level in the organizational
hierarchy who can help them with project
problems related to decisions about changes
in priorities and resource allocation.
Therefore, it is better if potential champions
belong to steering committees or project
portfolio advisory councils, or committees,
where issues related to project selection,
project priorities, and allocation of scarce
resources are discussed and decided. Then,
the project manager can influence the
right people already in the right place in
the organizational hierarchy to help solve
problems related to priorities and resources.

3. Why should they become "your" champion?

    To convince potential champions to become
    your real, strong champions, project
    managers must realize that most people
    usually do what is in their best interest. They
    must ask themselves why someone should
    become their champion, that is, What's in It

for Them (WIIFT). When project managers approach and request them to be their champion, they should describe the project outcomes and benefits in a way that turns potential champions on and emphasizes the following points:

- How their project aligns with organizational goals and strategies

- What's in It for Them (WIIFT), that is, for the champions

  They must highlight project benefits in a way potential champions appreciate at a personal level and at the organizational level. The first point helps champions accept the role more readily because they work not only for their personal benefit and satisfaction, but also help meet organizational goals and objectives.

  The second point should be addressed by speaking the champions' language. For example, if the champion is from the business unit of global programs, project managers should outline the project benefits in financial terms, such as rate of return (ROR), market segmentations, global market share, and product differentiation and specialization.

4. How should you groom them to be good champions?

   After project managers convince someone with appropriate positional power to become their champion, the next question is how to develop good champions. Project managers should help their champions win the battle with other steering committee members and high-level decision-makers to get decisions about resources and priorities that favor their project to achieve desired results.

Let us address these issues systematically. So far, project managers have answered the following questions satisfactorily:

Question 1. Who are real champions, and what are their appropriate roles and requirements?

Answer 1: Must satisfy two roles discussed earlier.

Question 2: Where are the real potential champions?

Answer 2: The potential champions are among the steering committees and senior management level where important decisions are made.

Question 3: Why should they become your champion?

Answer 3: Remember What's in It for Them (WIIFT), and then explain the project benefits in their language, which will motivate them to be your champion and advocate for you.

After these three questions are answered, champions are willing to go to bat for the project managers. The project managers must then remember that no one wants to fight battles they might lose. At this stage, project managers must help their champions win the battle for them so they want to continue to be their champions in the future. Project managers must do all the preparatory work for their champions and provide the following details about the project issues for which they need support from their champions:

- Data and information about the issue

- All relevant and up-to-date facts

- Overall analysis supported by proper tables and graphs

Project managers should present all this information effectively in a few slides (5–6), which helps convey the message powerfully. Project managers must provide their champions everything they need to prepare a convincing and powerful presentation.

If project managers do not prepare properly for their champions and expect them to do it on their own from detailed reports and so on, they put an extra burden on their champions, and senior management often might not have time for such additional tasks. It is in the project managers' best interest to do this very carefully, thoroughly, and professionally so their champions can easily argue for the project managers' concerns and find an appropriate solution to remove roadblocks.

5. How do you maintain your champions?

To ensure support from their champions in the future, project managers must cultivate good relationships and thank them whenever they receive help or advice. Most people feel good when they are given importance and respect. This situation is win-win for the project managers and their champions, leading to better working relationships.

	Issues	Recommendations
1	Who are the real champions?	• Must have appropriate formal (positional) power. • Must have appropriate influencing skills and network power.
2	Where to find real champions?	• Real champions are in senior management, steering committees, and prioritization panels, and they are major decision makers.
3	Why should they become your champions?	• Answer What's In It For Them (WIIFT). • Explain project benefits in their language.
4	How should you groom them to be good champions?	• Provide precise and concise data and facts to highlight main issues powerfully.
5	How do you maintain your champions?	• Thank them when they advocate for you and cultivate good working relationships with them for the future.

Table 4.1. Major Issues in Developing Good Champions, with Recommendations

4.1.6. Inappropriate decision-making

Inappropriate decision-making often leads to negative politics when senior managers make decisions that focus more on personal goals than organizational goals. Many other managers feel frustrated and upset about this behavior and the senior managers' decisions. They believe it will waste organizational resources and lead to undesirable politics. Furthermore, they feel unable to take corrective action because either they do not have more formal power to overrule those decisions, or they believe it is not worth getting into politics.

However, they continue to feel strongly that the senior managers are wrong, and it is bad for the organization when a few people can get away with making decisions for their gains rather than for the organization's benefit. Trust and respect for those managers decrease, leading to lack of cooperation and teamwork. Project managers must understand and manage politics effectively to stop a few senior managers from pushing their self-interests rather than ignoring politics to help the organization.

Several issues related to projects and project management processes lead to politics: project selection and prioritization, scarce resources, the resource allocation process, external projects, lack of a "real" champion, and inappropriate decision-making. Project managers should anticipate such issues and act to avoid the negative impact of politics.

4.2. People Behavioral Issues

> *We must learn to live together as brothers*
> *or perish together as fools.*
> —Dr. Martin Luther King, Jr.

Besides project management issues leading to politics, people and their interests and behaviors cause both negative and positive politics. In projects and organizations, people come from different backgrounds, viewpoints, norms, values, and cultures, so many work situations lead to different expectations and agendas, and people get involved in politics to pursue their agenda. Sometimes, negative politics arise when people try to achieve what they want without analyzing and evaluating the impact on their coworkers and their feelings. The following are common behavioral issues that lead to politics:

4.2.1. Personal egos and selfishness

Personal egos and selfishness are possibly the greatest and most common causes of politics. People often have egos at the personal level and the professional level, so they indulge in corporate politics to advance themselves and their interests. They are not sincere and genuine in working as a team, and they want more credit than they deserve.

The following are issues related to egos and selfishness:

- Networking for personal interests —
  Networking power is an effective power.
  Project managers should develop this power
  to help their projects and team members.

148

They can use their network to find new ideas, processes, systems (templates and software, and so on), and solutions to project problems. However, self-centered managers might misuse their networking power. They might steal credit for good ideas taken from others in their network to make themselves look good.

- Senior management branding management styles and their strategies — Sometimes, senior executives (presidents and CEOs) have too much ego, and they want to make superficial and unnecessary changes. They do not analyze their changes' impacts and just enforce them, which might not lead to improvement in organizational performance. They make changes to brand their management style and strategies to leave a legacy because of their high ego.

Many managers believe such executives do this for purely political reasons to satisfy their egos and self-interests. They must spend significant time and organizational resources to implement these changes just to satisfy their egos. This behavior leads to inappropriate use of resources better used to meet the goals of desired programs and projects to meet organizational strategies. Many managers feel frustrated, but they cannot stop these changes because superiors with more formal or positional power ask them to implement them.

4.2.2. Hidden agendas

Often, people who feel insecure about their job and skills have ulterior motives to satisfy self-interests. They hide their agendas and do not communicate openly with their project stakeholders. Sometimes, they have their ideas, and they want to implement processes and systems that help them meet their objectives. They try to present their ideas and describe the benefits with a "smoke and mirrors show," that is, in a way to argue their ideas are better than others are, and they should be implemented. They can use data and information to put their spin on the results.

However, if management analyzes and evaluates options and plans these people present, they can identify the weaknesses in the arguments presented. Management should call out such people and discourage the behavior that otherwise leads to wrong decisions for the organization.

4.2.3. Power struggles

Many project managers get involved in power struggles and turf wars and increase their power base by hook or by crook. The *crook* part leads to negative politics, whereas the *hook* part relates to legitimate strategies to increase power. Increasing informal power is the best strategy. When project managers have insufficient power and don't use it properly, they face many difficulties and challenges in managing their projects. They might play negative politics to get their projects done or blame other project managers for playing politics, even if those project managers genuinely help projects by using their informal power effectively and legitimately.

Project managers must help their team members increase their informal power so, as a team, they have more power and can empower and cross-train one another. Both the project manager and team members require maturity in modesty and commitment to help others grow. Project managers can use various techniques to increase power. However, they must recognize that the power must be used effectively to minimize the negative impacts of politics.

4.2.4. Withholding information

People in many organizations withhold information to build their power base. These people typically feel insecure about their job, expertise, and skills. They believe they can have power over others by not sharing information and making others depend on them for information they need to do their job effectively. They think that by withholding information about their subject matter expertise, they increase their demand and importance in the organization. They work in a silo and do not share their experiences to benefit the organization, which causes lack of cooperation and teamwork and leads to more negative politics.

4.2.5. Favoritism

People have informal relationships and connections in organizations, but when people give rewards or nominate others for rewards based on favoritism, rather than on objective criteria, it leads to negative politics. Many people believe others play this game for mutual gratification. Even nominating for rewards and special recognitions requires documentation from others, which gives them more visibility than those people deserve. Then, others reciprocate this behavior by nominating those who helped them in the past, and the practice continues. Others find this undesirable, leading to bad impressions, lack of respect, and finally to a less cooperative work environment and negative politics.

4.2.6. Boredom

Some are not challenged enough at work, and they feel bored, in line with the adage, "an idle man's brain is the devil's workshop." So, they are too involved in gossiping about other departments and people. Many such people could not spend so much time gossiping about others if they were busy meeting challenges at work that require them to do more research and analysis about project management processes and issues.

Projects are done by people with different interests or personalities. People behavioral issues—egos, hidden agendas, power struggles, withholding information, favoritism, boredom—lead to negative politics and a lack of synergy. Management should discourage such negative behaviors to create a culture of more collaboration and teamwork.

4.3. Organizational and Management Issues

> *To handle yourself, use your head; to handle others, use your heart*
> —Eleanor Roosevelt

Politics are common in most organizations, departments, and projects. Often, politics create situations and circumstances that management must handle. Some situations are beyond the control of the organization and its management, whereas some are caused by inability to anticipate changes that include cultural aspects, economic factors, global competition, and technological changes.

To cope with such challenges, management must restructure the organization and develop new strategies to optimize limited resources to compete and move forward successfully. Management must understand and analyze cultural aspects and people issues to implement desired changes. This section describes major organizational and management issues that lead to politics, with ideas to minimize the negative impact of politics. The following are major issues:

4.3.1. Restructuring

Different factors influence management to make major changes in the overall organizational structure:

- Growth rate and market share — Management might feel the need to restructure because of significant changes in the business growth rate. The business might be growing too fast or too slow, resulting in a gain or loss of market share. Both situations put pressure on top management to make organizational changes to cope with the changing growth rate.

- Technology (new technology to compete with and replace obsolete technology) — Sometimes, technological changes significantly influence major changes in organizations and management structure. Current technology has to be abandoned because of technical obsolescence, and major investments have to be made in new technology to compete at the global level successfully. Organizations are forced to make major changes to optimize their staff's skill mixture to take full advantage of new technology.

- Staff shuffling — Decisions are made to hire more staff or lay off staff, and people involved become apprehensive because of many uncertainties and issues such as the following:

  - Will their roles and responsibilities be expanded or shrunk?

  - Will they be promoted, fired, or shifted laterally?

  - What will happen to their departments, supervisors, and allies?

    All this uncertainty and apprehension leads to political activities and behaviors by several managers to position themselves properly and protect their departments and favorite teams.

4.3.2. Cultural aspects

Project teams consist of people with different backgrounds, interests, and cultures. Cultural aspects are important because they represent a set of shared values and beliefs. People with different cultures do things differently because of their different beliefs, viewpoints, and orientations. However, some project personnel misinterpret these different ways as politics rather than merely cultural differences.

4.3.3. Change management

Change is a fact. It affects most business organizations and projects (Doyle 1990, 36–45; Offerman & Gowing 1990, 95–108).[1] In today's business environment, these changes are caused by rapid changes in technology and global competition. Project managers must manage their projects in a virtual team environment. Corporate structures and project management strategies change continually because of globalization of markets, joint ventures, downsizing, and flattening of organizations Kanter 1991, 151–164).[2]

In today's business world with economic and labor constraints, MBP is becoming an effective way to manage organizations. Projects create changes, and project managers are recognized as change agents. Organizations' success and failure depend on their ability to manage change successfully. Change causes conflicts, stress, and insecurity because the outcomes of change are unpredictable and often unknown. Therefore, many people resist change and call the motives to change and change management process politics.

To develop and implement strategies to manage change effectively, managers must understand the dynamics and principles of change (Verma 1997, 18–20).[3] Managers must understand the reasons for and sources of resistance to change (Hellriegel, Slocum & Woodman 1992, 719–742).[4, 5] One common reason for resistance to change is that people are stressed because of the unknown outcomes. Therefore, managers must listen to their people's concerns first, and then explain the reasons for change and potential benefits of change. They must analyze major reasons (personnel and others) behind the resistance to change and develop strategies to overcome the resistance to change (Hellriegel, Slocum & Woodman 1992, 719–742; *Management Review* 1972, 17–25).[6]

Kurt Lewin suggested a Force Field Analysis approach to overcoming resistance to change. This analysis is because introducing change requires balancing of opposing forces and reaching equilibrium (1951; *Human Relations* 1951, 5–41).[7] An effective approach to overcome resistance to change and manage change is to implement changes in small steps and let people experience the benefits of change for themselves and their projects. Then, they do not consider the changes and change process politics, and they will be more open to new ideas to change. They will even likely cooperate in implementing the change management process.

4.3.4. Tolerance for negative politics

One reason for continuing negative politics in most organizations is that senior management tolerates negative politics. If senior management made a "zero tolerance" policy for negative politics, just as with drugs at work and school, people would be more cautious in involving themselves in negative politics. With this organizational policy, senior management would discourage project managers from playing negative politics and urge them to take steps to create a culture of more collaboration, openness, and cooperation. Management should promote positive politics by following attributes and ideas for positive politics and minimizing the negative impacts of politics.

4.3.5. Reward system

A proper reward system based on objective criteria reduces the likelihood of undesired politics. A subjective reward system based on favoritism leads to anxiety and frustration among project personnel and to negative politics. In addition, it lowers project personnel's trust in senior management.

Instead, management should develop objective criteria for giving rewards and recognition to higher performers in consultation with project managers, team members, and other stakeholders. When project personnel are involved, they are more likely to accept reward systems as fair and cooperate in implementing them.

Organizational and management issues—restructuring, cultural aspects, change management, tolerance for negative politics, subjective reward systems—often lead to politics. Management should address these issues promptly to reduce the negative impact of politics.

4.4. External and Global Market Environment Issues

> *It is not the strongest of the species that survive, nor the most intelligent, but the one most responsive to change.*
> —Charles Darwin

Generally, people cause politics. However, sometimes, the underlying reason is external factors beyond their control. When it happens, people often become anxious because they cannot predict what will happen to their existing positions, roles, and responsibilities. The following are external factors:

4.4.1. Mergers, acquisitions, and joint ventures

Mergers, acquisitions, and joint ventures are becoming common because of globalization and pressure to increase profits and share prices and bridge the gap in organizational strength and capabilities. Organizations respond to such challenges by merging with other compatible organizations, acquiring new companies, or collaborating with other businesses worldwide to increase their market share and profits. All these factors might lead to internal changes in the organizational culture and, therefore, significant changes in communication styles, decision-making styles, and so on. Again, these endeavors create uncertainties and increase anxiety and apprehension among people who then play politics to position themselves appropriately to survive and possibly advance in the future.

4.4.2. Regulations

Often, changes in regulations are decided and imposed by the lawmakers of the nation or regions where the project is done. Nevertheless, they can have serious impacts on how parent companies organize their projects in different countries to cope with these regulations.

These changes might be about privatization or putting organizations under government control. Often, such changes are made primarily for political reasons and cause internal organizational politics. They might lead to significant changes in the workload for people and departments to satisfy new rules and regulations. There might be considerable changes in policies and procedures associated with several areas: design reviews, manufacturing, inspection, quality assurance procedures, and purchasing policies, documentation requirements, and training policies. They might add or shift workload in departments. Again, these changes lead to anxiety among people involved because of the following uncertainties:

- What organizational changes will be made

- How their positions, roles, and responsibilities will be affected

Global market environments, including mergers, acquisitions, and joint ventures, and regulations imposed by governments and businesses create significant anxiety and politics. Some are unavoidable and uncontrollable , but management should be proactive and act appropriately to address the issues and minimize the negative impact of politics.

Chapter 4 Summary

Globalization of markets and business operations can cause politics in organizations. Many issues related to such factors are external because they are beyond the control of project managers and team members. Nevertheless, project personnel become anxious because of uncertainties about their future roles, responsibilities, and power base. Such issues include mergers, acquisitions, joint ventures to combine expertise and resources, and changes in regulations (such as privatization or government controlled) that change the nature of jobs, roles, and responsibilities of project managers, project teams, and project management processes.

Politics are unavoidable in most organizations. The reasons for politics can be broadly divided into four categories:

1. Projects and project management issues

    Projects are done by people, and wherever there are people, there are politics. Projects have stakeholders with different attitudes, expectations, behaviors, and personalities. Issues related to projects and project management include the following:

    *   Project selection and prioritization

    *   Scarce resources

    *   Resource allocation process

    *   External projects

    *   Lack of "real" champion — A real champion has two roles: person with a vested interest and remover of roadblocks.

Project managers often ask the following five questions about champions:

1. Who are the real champions?
2. Where are the real champions?
3. Why should they become "your" champion?
4. How do you develop good champions?
5. How do you maintain your champions?

- Inappropriate decision-making

2. People behavioral issues

   Project managers should try to understand various stakeholders' attitudes and behaviors and their motives behind them to minimize the negative impact of politics. People behavioral issues include the following:

   - Personal egos and selfishness — Egos and selfishness create the following issues: networking for personal interests and senior management branding management styles and their strategies.
   - Hidden agendas
   - Power struggles
   - Withholding information
   - Favoritism
   - Boredom

3. Organizational and management issues

   Senior management should be proactive in anticipating such issues and challenges to curtail the impact of politics. These issues and challenges include the following:

   - Restructuring — Changes in the overall organizational structure can occur because of growth rate and market share, technology, and staff shuffling.
   - Cultural aspects
   - Change management
   - Tolerance for negative politics
   - Reward system

4. External and global market environment issues

   Organizational leadership teams should constantly evaluate global market situations and develop strategies to resolve such issues and challenges. These issues lead to anxiety and politics and include the following:

   - Mergers, acquisitions, and joint ventures
   - Regulations

Politics are inevitable in project environments. However, project managers must understand the dynamics of politics and the main reasons and issues leading to politics and develop skills to minimize the negative impact of politics.

Politics in Managing Stakeholders

Pay special attention to powerful stakeholders.

—Anonymous

Politics are common in most organizations and, especially, in project environments because projects have stakeholders with different interests, beliefs, personalities, and skills. According to the *PMBOK Guide* (2013, 391–397), developed by PMI, *stakeholder* can be defined as any individual, group, or organization who may affect, be affected by, or perceive to be affected by a decision, activity, or outcome of the project, program, or portfolio.[1]

All stakeholders have their viewpoints about how things should be done, especially about the impact of politics to get things done in project environments. Stakeholder management includes the processes required to identify stakeholders, analyze stakeholders' expectations and their impact on the project, and to develop appropriate management strategies for effectively engaging stakeholders in project decisions and execution (*PMBOK Guide* 2013, 391–397).[2]

Politics and projects are inseparable. Often, projects create change that could be positive or negative for people. Change creates uncertainty and anxiety and, therefore, leads to conflict and stress. In project environments, there are conflict and competition for limited resources.

Other issues that lead to disagreements and conflicts include allocation of power and responsibilities, budget allocation, salaries, and bonuses. Project stakeholders use power and politics to get what they want in such uncertain or conflict situations. Research indicates that only 25 percent of stakeholders who receive the changes intended from project completion commit to specific changes, whereas the remaining 75 percent resist or follow the changes with the breakdown—21 percent (resist) and 54 percent (comply) (*Changefirst* 2014).[3]

5.1. Politics and Stakeholders

> *Every stakeholder likes to do what he likes*
> *and not what should be done.*
>
> —Anonymous

Project stakeholders often have diverse personalities, viewpoints, and backgrounds. Key stakeholders can make or break a project's success. Even if project managers meet schedule, cost, and quality requirements, no one is happy if the stakeholders are unhappy. Therefore, effective stakeholder management is the key to project success. It requires political skills to understand the power structures and motives of different stakeholders.

5.1.1. Importance of politics in managing stakeholders

> *[The company does] a lot of things for reasons besides profit motive. We want to leave the world better than we found it.*
> —Tim Cook, CEO of Apple

Lynda Bourne and Derek Walker (2004, 226–243) suggested that relationship management helps address stakeholders' expectations throughout the project lifecycle. They described the concept of a set of skills— *tapping into the power lines*—to understand the power source that drives projects and programs in organizations and learn techniques to harness this energy to deliver successful projects.[4]

Project outcomes are influenced by hidden agendas and overt actions of various people and groups. In large organizations, politics can be defined as understanding the power structures and using them to influence project outcomes. Project managers should know their project stakeholders' needs, interests, and expectations and develop skills to manage different categories of stakeholders and their concerns to deliver successful projects.

Crawford and Da Ros (2002, 20.1–20.10) conducted quantitative (questionnaires) and qualitative (interviews) research to study the impact of organizational politics on project outcomes and the importance of political skills for managing projects successfully. Their studies indicated the following two points:[5]

1. Organizational politics and acquisition of project resources are highly correlated.

2. Project managers' ability to manage organizational politics is crucial to project success.

Jeff Pinto (2000, 85–91)) suggested that politics are a natural way to get things done in project environments, and therefore, project managers must develop political skills to get things done through others. He suggested that a political behavior to seek, acquire, and maintain power is pervasive in modern corporations.[6]

Politics are common in most project environments because of diverse stakeholders. All stakeholders have their viewpoints about how things should be done in an organization. Often, there are significant politics when their viewpoints are not only different, but they also have strong enough conviction in their viewpoints to fight hard to push their ideas. To manage stakeholders, project managers must understand stakeholders' political behaviors and the political dynamics among stakeholders.

One key to successful stakeholder management from a political viewpoint is to have the "right" project champion with enough organizational influence to remove project obstacles related to politics, lack of resources, and support from senior management, to achieve project objectives. Project managers should be politically savvy, and they must have combined influencing and leadership skills to gain commitment of important stakeholders.

In the *PMBOK Guide*, the closest word to "champion" is a "sponsor." However, sponsors might not be good champions. They could be different champions during different phases of the project lifecycle, especially when a particular project introduces a major change in the organization related to organizational structure, policies, and processes involving the selection and prioritization of projects and allocation of people and financial resources.

According to PMI literature (2014, 2013), effective sponsorship is a key to delivering successful change projects.[7] No one person is the perfect sponsor throughout the project lifecycle. Harrington and Nelson (2013) developed a sponsor evaluation assessment checklist that can be used to select sponsors and measure the effectiveness of their performance.[8] Project managers can use this checklist to develop common understanding with their sponsors that they would fulfill specific behavioral roles effectively and demonstrate their conviction by appropriate actions This helps their sponsors be effective champions for removing most project obstacles, as required, to deliver successful projects.

From a political viewpoint, effective stakeholder engagement is the key to successful stakeholder management. According to Miller and Oliver (2015, 1–23), stakeholder management must focus on active engagement of stakeholders throughout the project lifecycle.[9] Project managers must understand the dynamics of politics in their organizations while managing stakeholders. They must realize that stakeholders and their relationships with one another are likely to change as a project progresses through its life cycle, which influences how project managers should deal with the politics related to stakeholder management. The following are common questions from many project managers about political dynamics among stakeholders and managing stakeholders:

- Who are major project stakeholders, and what do they think and why?

- What influence do they have on the project?

- How do we communicate with all stakeholders?

- How do we engage them and gain their commitment?

For effective management of stakeholders, project managers must understand the main aims and goals of stakeholder management. In this context, project managers identify stakeholders and assess how they are likely to influence the project or how the project might affect them. The goal of project managers is to develop cooperation between stakeholders and the project team and ultimately ensure project success by gaining commitment and teamwork from all project stakeholders.

5.1.2. Identifying stakeholders

If you work for and eventually lead a company, understand that companies have multiple stakeholders including employees, customers, business partners and the communities within which they operate.
—Don Tapscott

An appropriate and thorough identification of stakeholders is the primary step in understanding political dynamics and relationships for managing stakeholders. Major stakeholder categories can be described as follows (Verma 1995, 46–70):[10]

- **Core stakeholders** — These stakeholders are intimately involved in the project daily, and they are continuously concerned about project activities, problems, and outcomes. Core stakeholders are the project manager, project team, client, and project sponsor.

- **Internal stakeholders** — These stakeholders are directly involved with the project. Major internal stakeholders are top management, functional managers, resource managers, organizational staff and service personnel, and other project managers.

- **External stakeholders** — These stakeholders are indirectly involved with the project, and they can indirectly influence project outcomes. External stakeholders include regulatory agencies and personnel at regional, state, and federal levels; competitors; environmental and legal agencies; public and press; social and cultural organizations; economic and financial organizations; and various professional organizations and vested groups. Contractors, subcontractors, and major vendors can be identified as internal or external stakeholders.

- **Unexpected stakeholders** — Project managers are usually unaware of these stakeholders. These stakeholders can have a positive or a negative impact, and they can be identified as internal or external. These stakeholders should be identified as early as possible in the project lifecycle and managed appropriately. Tres Roeder (2013, 97–101) described these stakeholders as "phantom stakeholders" and suggested techniques to identify them by looking at unexpected changes and major missing items in the project.[11]

Project managers must identify stakeholders from all these categories and then understand and analyze political dynamics among the stakeholders to develop strategies to manage them effectively.

There are always politics in project environments because of a diverse mix of stakeholders with different roles, interests, and personalities. Project managers must identify different categories of stakeholders (core, internal, external, and unexpected) and understand the importance of managing them, keeping in mind that effective stakeholder engagement is the key to successful stakeholder management.

5.2. Analyzing Stakeholders

To make progress we have to build a multi-stakeholder process, harnessing the appropriate energies.
—Mary Robinson

After identifying stakeholders, the next step is to analyze them for their interests, needs, strengths, and weaknesses. Stakeholder analysis should be done continually throughout the project lifecycle. *Stakeholder analysis* is systematically gathering and analyzing qualitative information to determine whose interests should be taken into account for how much and how seriously when developing and implementing a program or a project.

Stakeholder interests affect project riskiness and viability. Project managers should understand and clarify stakeholder interests, needs, and capabilities and identify potential opportunities, risks, and threats. They should also determine the degree of involvement for certain groups of stakeholders in project planning, implementation, and evaluation.

The following are reasons for doing stakeholder analysis:

- To understand goals and interests

- To analyze political structure

5.2.1. Understanding goals and interests

> *Talk in terms of other person's interests.*
> —Dale Carnegie

Understanding stakeholder goals and interests is critical to stakeholder analysis. Adrienne Watt (2010, 1–10) identified the following stakeholder goals and interests:[12]

- What do they want (what are the real driving forces)?

- What is each after (their real potential gains or losses)?

171

- Who is most important to be looked after among all stakeholders?

- Who has the most authority to change project plans and execution strategies?

- Are there hidden agendas or goals stakeholders push?

- What are the goals of stakeholders with high power and the ability to have the greatest impact? To understand stakeholder goals and interests, project managers should identify the real problems by raising the questions, "what is the underlying problem?" and "what are the differences between the motives and expressed expectations of stakeholders?" The following questions should be asked when trying to understand stakeholder goals and interests:

- Have all stakeholders been listed? This question helps identify all stakeholder expectations and benefits they seek from project outcomes.

- Have all potential supporters and opponents been listed? This question's purpose is to identify what resources particular stakeholders would commit or avoid committing to the project.

- Are new stakeholder groups likely to emerge as a project outcome? This question helps identify other interests stakeholders have that might conflict with project objectives. This question also sheds light on how much they respect other stakeholders on the list.

Project managers should understand stakeholder goals and interests to determine who the real supporters are and who are likely to introduce obstacles to project success. This understanding helps project managers analyze stakeholders' political behaviors and develop strategies to manage their behaviors to avoid negative impact on projects.

5.2.2. Analyzing political structure

The word is stakeholding. The style is integrity. The profession is business.
—Anita Roddick

To understand political dynamics among stakeholders, you must analyze the organizational structure from a political viewpoint. Analyzing the political structure helps recognize informal relationships critical in project execution and implementing required changes.

As a first step, project managers should identify main characteristics and political features of various stakeholders' roles and identify stakeholders associated with these characteristics, with an associated symbol on the organization chart. However, they should be cautious to keep it confidential rather than openly sharing with everyone. The following are stakeholder characteristics you should identify to analyze the political context around the project:

- Who are the true allies and project champions?
- Who are real supporters (range can be from medium to high)?

- Who are knowledgeable about the project and organizational policies and procedures to know the shortcuts and possible deviations?

- Who are power brokers, shakers, and movers in the organization?

- Who are likely to win or lose the most from the project's success or failure?

- Who has the most formal authority to make changes?

- Who can most influence the important decision-makers and managers to get extra financial and human resources, as needed, to push the project?

- Who is readily accessible to discuss project strategies, issues, and problems?

- Who is likely to sabotage the project?

Project managers must be extra careful of such stakeholders because they might go to any length to harm them and the project.

Organization structures typically identify formal relationships, but you must also understand informal relationships and political dynamics among stakeholders throughout the organization. Therefore, project managers must analyze the political structure by understanding stakeholder characteristics. This analysis is necessary to understand who has the real power, most interest, and high influence to make major changes to project plans, execution strategies, and project outcomes. This helps you understand and manage stakeholders' politics.

Analyzing stakeholders for their interests, needs, strengths, and weaknesses is the key to managing them and the associated politics. Project managers must understand stakeholder goals and interests and analyze the political structure to manage politics effectively, which involves identifying stakeholder characteristics to analyze the political context around the project.

5.3. Prioritizing Stakeholders Using Main Attributes

To gain stakeholders' cooperation, it is important to find out what they need, when they need, why they need, and how they need.
—Anonymous

The purpose of analyzing stakeholders is to prioritize them by who are most important based on their power level and other attributes. This section identifies relationships among levels of authority (formal power) and various attributes of all stakeholders and describes how project managers should communicate with their stakeholders and manage them effectively to deliver successful projects.

The attributes can be divided into the following two categories:

Category 1: Attributes related to stakeholder interests (Roeder 2013, 97–101; *PMBOK Guide* 2013, 395–397):[13]

1. Project knowledge (subject matter expertise and knowledge about the project objectives, planning, and execution strategies)

2. Interest (level of concern)

3. Influence (level of active involvement)

4. Impact (ability to effect changes to project planning or execution)

Category 2: Attributes related to stakeholder traits:

1. Predictability (how predictable stakeholders are under various situations)

2. Commitment to change (personal convictions about the project benefits)

3. Respect (level of respect by most people in the organization)

4. Ethical standards (moral philosophy that defines right or wrong)

5. Integrity (honesty, trustworthiness, morality, and uprightness of character)

5.3.1. Category 1: Attributes related to stakeholder interest

Find the appropriate balance of competing claims by various groups of stakeholders. All claims deserve consideration, but some claims are more important than others.
—Warren Bennis

Table 5.1 shows matrices according to power levels mapped with levels of first category of stakeholder attributes to prioritize them and then develop communication strategies to manage them (Roeder 2013, 97–101; *PMBOK Guide* 2013, 395–397).[14]

Attributes		Low Power	High Power
Knowledge (K) (subject matter)	H	Keep Informed (capitalize on knowledge and experience)	Strengthen Working Relationships (gain support and knowledge)
	L	Monitor (reduce risk of wrong actions)	Keep Satisfied (avoid negative influence)
Interest (C) (level of concern)	H	Keep Informed (avoid negative emotional responses)	Keep them Involved and Interested (possible allies)
	L	Monitor (minimal effort to ensure positive actions)	Keep Satisfied (update about project benefits)
Influence (I) (level of active involvement)	H	Keep Informed (empower for better performance)	Recognize Efforts and Support (work closely)
	L	Monitor (minimal effort)	Keep Satisfied (update information to increase interest)
Impact (Im) (ability to effect changes to planning & execution)	H	Keep Informed (gain cooperation)	Strengthen Working Relationships (show what's in it for them)
	L	Monitor (minimal effort)	Keep Satisfied (gain formal support)

Table 5.1. Matrices Showing Levels of Power and Category-1 Attributes of Stakeholders

1. Power versus knowledge matrix

Knowledge in the matrix shown in table 5.1 refers to project-specific knowledge. In addition, knowledge can include subject matter expertise about the project that management and other stakeholders across the organization recognize and respect. Management shows high confidence in the stakeholders' knowledge and experience. Stakeholders with high knowledge are likely to contribute more to the project with their ideas and experiences. Knowledgeable stakeholders can also help execute project activities according to project plans and objectives.

As shown in table 5.1, when power levels are mapped with knowledge levels, stakeholders can be classified into four quadrants on the power/knowledge matrix, and the following are associated strategies project managers can use to manage stakeholders:

- **Low Power/High Knowledge (LP/HK): Keep Informed (Capitalize on knowledge and experience)**

  Stakeholders in this quadrant have high knowledge, but do not have enough power to make changes. Therefore, project managers should keep these stakeholders informed and capitalize on their high knowledge and experience.

- **Low Power/Low Knowledge (LP/LK): Monitor (Reduce risk of wrong actions)**

  These stakeholders have low power and low project knowledge, and therefore, project managers should monitor them

178

continually to get desired cooperation, as needed, and minimize wrong actions.

- **High Power/High Knowledge (HP/HK): Strengthen Working Relationships (Gain support and knowledge)**

  The high level of stakeholder project knowledge in this quadrant shows that the current communication plan is working well, and project managers should continue it. Project managers should establish and strengthen good working relationships with these stakeholders and capitalize on their high power and knowledge to gain formal support for their project and develop sound project management strategies.

- **High Power/Low Knowledge (HP/ LK): Keep Satisfied (Avoid negative influence)**

  These stakeholders have high power and might cause disruptions because of their lack of project knowledge. Therefore, project managers should give high priority to communicating with these stakeholders regularly to keep them satisfied and avoid risk of negative influence on the project.

2. **Power versus interest matrix**

   Interest on the matrix in table 5.1 refers to stakeholders' concern for project outcomes. The more concerned they are about the project, the more interest they take during project planning and execution. Depending on their power level, they are motivated to act to affect the project outcomes to satisfy themselves. The four quadrants in this matrix can be described as follows:

- **Low Power/High Interest (LP/HC): Keep Informed (Avoid negative emotional responses)**

  Stakeholders in this quadrant have high concern about the project but a low power level. Therefore, project managers must keep these stakeholders informed by giving them relevant information at the right time, without spending too much time, to avoid negative emotional responses.

- **Low Power/Low Interest (LP/LC): Monitor (Minimal effort to ensure positive actions)**

  This quadrant includes stakeholders with low power and low interest. Therefore, project managers should spend minimum effort dealing with these stakeholders.

- **High Power/High Interest (HP/HC): Keep Them Involved and Interested (Possible allies)**

  Stakeholders in this quadrant are important because they have high power and interest in project outcomes. Project managers should manage them closely by involving them early in the project and keeping them up to date continually to maintain interest. Politically, project managers should have stakeholders from this quadrant as their allies and champions.

- **High Power/Low Interest (HP/LC): Keep Satisfied (Update about project benefits)**

  This quadrant includes stakeholders with high power and low interest in project outcomes. As a communications strategy, project managers should handle them carefully and keep them satisfied by giving them required information promptly to ensure they support the project. These stakeholders could be powerful if they take more interest in the project outcome and move to the quadrant with high power/high interest.

3. **Power versus influence matrix**

   Power and *influence* might seem similar. However, *power* here refers to the authority or formal power that might lead to resistance and might help get compliance, whereas influence here is inclined more toward the informal side of power that helps get commitment. *Influence* refers to the levels of stakeholders' active involvement in the project to influence project execution strategies and outcomes. The four quadrants in this matrix can be described as follows:

- **Low Power/High Influence (LP/HI): Keep Informed (Empower for better performance)**

Stakeholders in this quadrant have low power and high influence, meaning they are more involved but lack power to make changes. Project managers should keep these stakeholders informed to empower them for better performance and take advantage of their high influence whenever they can.

- **Low Power/Low Influence (LP/LI): Monitor (Minimal effort)**

  Project managers should spend the least effort managing stakeholders in this quadrant because they have low power, and they are not involved in the project.

- **High Power/High Influence (HP/HI): Recognize Efforts and Support (Work closely)**

  These stakeholders are important, and project managers must work closely with them and recognize their efforts and support because they have high formal power. They are involved in the project to change project plans, execution strategies, and outcomes.

- **High Power/Low Influence (HP/LI): Keep Satisfied (Update information to increase interest)**

  Stakeholders in this quadrant have high power, but they might not be involved in your project during a particular time because they are too busy with other projects and their functional, or line, responsibilities during that time.

Project managers must not only keep these stakeholders satisfied and updated as a minimum, but also manage these stakeholders closely because they might decide to be more involved in project execution and key decisions as the project progresses to a phase of their interest.

4. **Power versus impact matrix**

   Impact refers to the stakeholder's ability to change project plans and execution strategies. Impact differs from influence (level of active involvement) and interest (level of concern) in that stakeholders might have high influence and interest, but they cannot change project planning and execution. The four quadrants in this matrix can be described as follows:

   • **Low Power/High Impact (LP/HIm): Keep Informed (gain cooperation)**

     Stakeholders in this quadrant might have greater ability to effect changes, but not have enough power to implement them. Therefore, project managers should keep these stakeholders informed to gain cooperation and be ready to take advantage when they can effect positive change.

   • **Low Power/Low Impact (LP/LIm): Monitor (Minimal effort)**

     Project managers should monitor stakeholders in this quadrant and spend minimal effort in managing them because of their low power and ability to make changes.

- **High Power/High Impact (HP/HIm):
  Strengthen Working Relationships
  (Show What's in It for Them (WIIFT))**

  Stakeholders in this quadrant are
  important because of their high power and
  ability to make changes. Project managers
  should establish and strengthen good
  working relationships with these
  stakeholders and convince them to act as
  allies and champions by showing What's
  in It for Them (WIIFT).

- **High Power/Low Impact (HP/LIm):
  Keep Satisfied (Gain formal support)**

  Stakeholders in this quadrant might
  have less ability to make changes, but
  because of their high power, they might
  influence others with high power to make
  changes undesirable for project managers.
  Therefore, project managers should keep
  these stakeholders satisfied by giving
  them relevant information promptly to
  gain their formal support.

After stakeholder analysis is done, project
managers should prioritize stakeholders by mapping
their levels of power and Category-1 attributes that
include knowledge, interest (level of concern), influence (level of active involvement in the project), and
impact (ability to effect changes to project planning
and execution) to develop strategies to manage
stakeholders.

5 - Politics in Managing Stakeholders

5.3.2. Category 2: Attributes related to stakeholder traits

To effectively communicate, we must realize that we are different in the way we perceive the world and use this understanding as a guide to our communication with others.
—Anthony Robbins

Figures 5.1–5.5 show matrices according to power levels mapped with levels of second category of stakeholder attributes related to stakeholders' personal traits to prioritize them and then develop communication strategies to manage them. The matrices are explained as follows:

1. **Power versus predictability matrix**

   Robert Newcombe (2003, 841–848) described a stakeholder-mapping approach to analyze and manage stakeholders for construction and development projects, and suggested a matrix showing this relationship.[15]

   Figure 5.1 refers to how predictable various stakeholders are and shows the relationship between levels of power and *predictability* of various stakeholders. As shown in figure 5.1, when power levels are mapped with levels of predictability, stakeholders can be classified into four quadrants that show associated strategies to deal with stakeholders on the power/predictability matrix.

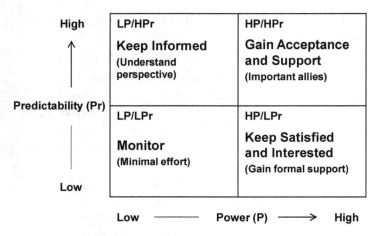

High ↑	LP/HPr **Keep Informed** (Understand perspective)	HP/HPr **Gain Acceptance** **and Support** (Important allies)
Predictability (Pr) Low	LP/LPr **Monitor** (Minimal effort)	HP/LPr **Keep Satisfied** **and Interested** (Gain formal support)

Low ——— Power (P) ——→ High

Figure 5.1. Matrix showing levels of power and predictability.

The four quadrants in this matrix can be described as follows:

- **Low Power/High Predictability (LP/HPr): Keep Informed (Understand perspective)**

  These stakeholders might present a few problems, and therefore, project managers should keep them informed. Project managers should stay in touch with their stakeholders and understand where they are coming from and why.

- **Low Power/Low Predictability (LP/LPr): Monitor (Minimal effort)**

  These stakeholders are unpredictable, but they are easily manageable because of their low power.

- **High Power/High Predictability (HP/ HPr): Gain Acceptance and Support (Important allies)**

  Stakeholders in this quadrant are important because of their high power and predictability. Project managers must manage them closely and gain their acceptance and support for the project. These stakeholders should help project managers implement desired changes because of their high formal power.

- **High Power/Low Predictability (HP/ LPr): Keep Satisfied and Interested (Gain formal support).**

  Project managers should not ignore these stakeholders, and they must keep them satisfied and interested because the stakeholders have low predictability but high formal power. These stakeholders can pose opportunities and difficult challenges. Project managers should use the help of stakeholders with high power and high predictability to influence stakeholders in this category to gain their support.

2. **Power versus commitment to change matrix**

    David Miller and Mike Oliver (2015) described a matrix showing the relationship among levels of power and levels of *commitment to change*.[16] Commitment to change differs from impact (an ability to change) in that *impact* implies an ability to change, whereas *commitment to change* refers to the level of deep interest in the project and strong conviction to make desired changes. Figure 5.2 shows strategies to deal with stakeholders at varying levels of power and commitment to change.

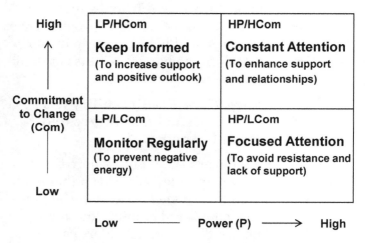

High ↑ **Commitment to Change (Com)** ↑ **Low**	**LP/HCom** **Keep Informed** (To increase support and positive outlook)	**HP/HCom** **Constant Attention** (To enhance support and relationships)
	LP/LCom **Monitor Regularly** (To prevent negative energy)	**HP/LCom** **Focused Attention** (To avoid resistance and lack of support)

Low ——— **Power (P)** ——→ **High**

Figure 5.2. Matrix showing levels of power and commitment to change.

As shown in figure 5.2, when power levels are mapped with levels of commitment to change, stakeholders can be classified into four quadrants that show associated strategies to deal with stakeholders on the power/commitment to change matrix. The four quadrants in this matrix can be described as follows:

- **Low Power/High Commitment (LP/ HCom): Keep Informed (To increase support and positive outlook)**

Project managers should gain these stakeholders' support without spending too much energy because of their low power to push the project. However, project managers should keep these stakeholders informed and share with them early good news about the project so they stay positive and supportive of the project.

• **Low Power/Low Commitment (LP/ LCom): Monitor Regularly (To prevent negative energy)**

Project managers should monitor stakeholders regularly in this quadrant to prevent negative energy. The project manager and project team should be careful about these stakeholders because if they feel ignored, they can influence more powerful stakeholders negatively.

• **High Power/High Commitment (HP/ HCom): Constant Attention (To enhance support and relationships)**

Stakeholders with high power and commitment to change are valuable and need constant attention to enhance support and relationships. Project champions can be in this quadrant, and project managers must ensure they have their constant support and make the support visible to others in the organization. Project managers should involve them early in the project and show their gratitude for their support and cooperation.

- **High Power/Low Commitment (HP/ LCom): Focused Attention (To avoid resistance and lack of support)**

  Stakeholders in this quadrant need focused attention from the project manager and the project team to avoid resistance and lack of support. The project manager and project team should watch these stakeholders, because if they feel ignored, they can be more resistant, spread negative energy, and use their power to influence project activities negatively.

3. **Power versus respect matrix**

   The attribute of *respect* refers to the level of earned respect from most people in the organization. Some senior management members have high power, but not high respect. Figure 5.3 shows strategies to deal with stakeholders at varying levels of power and respect.

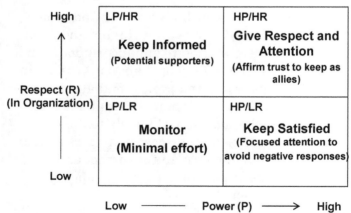

Figure 5.3. *Matrix showing levels of power and respect.*

As shown in figure 5.3, when power
levels are mapped with levels of respect,
stakeholders can be classified into four
quadrants that show associated strategies to
deal with stakeholders on the power/respect
matrix. The following are recommendations
for managing stakeholders in the four
quadrants of this matrix:

- **Low Power/High Respect (LP/HR):
  Keep Informed (Potential supporters)**

  Project managers should keep these
  stakeholders informed, because even
  though they have less power, they have
  high respect. If these stakeholders feel
  ignored, they can influence stakeholders
  with high power negatively. Additionally,
  if they are well informed, they can be
  potential allies when they gain more
  formal power.

- **Low Power/Low Respect (LP/LR):
  Monitor (Minimal effort)**

  Stakeholders in this quadrant should be
  monitored so they don't create negative
  energy among other stakeholders. Project
  managers should spend minimal effort in
  managing them.

- **High Power/High Respect (HP/HR):
  Give Respect and Attention (Affirm
  trust and agreement to keep as allies)**

  These stakeholders are like project
  champions, and they are valuable. Project
  managers should manage them closely
  and give them due respect and attention.

Project managers should reaffirm
the trust and agreement with them
continually.

- **High Power/Low Respect (HP/LR):
  Keep Satisfied (Focused attention to
  avoid negative responses)**

  Project managers should keep stakeholders in this quadrant satisfied and give
  them focused attention to avoid negative
  responses. Project managers should stay
  in touch with them because of their high
  power and occasionally point out to them
  the project benefits for the organization
  and for their specific department to
  sustain their support.

4. **Power versus ethics matrix**

   Ethics refers to a philosophy or code of morals
   that defines right or wrong. This code of
   morals is often called a *code* of *ethics* that
   refers to standards of character set up by any
   race, nation, or organization. Being ethical
   means acting in accordance with correct
   principles defined by a given system of ethics
   and professional conduct in organizations,
   associations, groups, and teams. Stakeholders
   with high levels of ethical standards are
   unafraid to do and say the right things under
   any circumstances. They are most likely
   to follow the organizational policies and
   procedures and not abuse their powers as a
   higher priority than their benefits.

As shown in figure 5.4, when power levels
are mapped with levels of ethics, stakeholders
can be classified into four quadrants that
show associated strategies to deal with stakeholders on the power/ethics matrix.

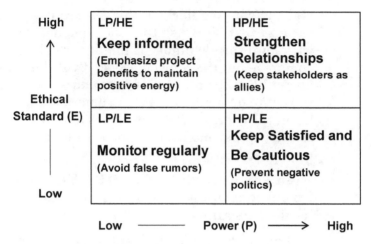

Figure 5.4. Matrix showing levels of power and ethics.

The following are associated strategies project
managers can use to manage stakeholders:

- **Low Power/High Ethics (LP/HE): Keep
  Informed (Emphasize project benefits to
  maintain positive energy)**

  Stakeholders in this quadrant have low
  power but high levels of ethical standards,
  and therefore, project managers should keep
  them informed about and emphasize project
  benefits so they stay positive and minimize
  negative impacts of politics about the project,

even if they cannot influence major changes because of their low power.

- **Low Power/Low Ethics (LP/LE): Monitor Regularly (Avoid false rumors)**

  Project managers should monitor stakeholders regularly in this quadrant because there is a risk that the stakeholders with low ethical standards might spread false rumors about the project, especially if they do not like the project manager and some project team members. False rumors might harm the overall project progress and even relationships among stakeholders and project team members.

- **High Power/High Ethics (HP/HE): Strengthen Relationships (Keep stakeholders as allies)**

  Stakeholders in this quadrant are important, and project managers must manage them closely. They say and do the right things because of their high levels of ethics, and because of their high level of power, they can help the project manager get resources and resolve project problems to meet project objectives. Project managers must keep them informed about the project benefits and successes in executing project plans and give them enough respect and recognition to strengthen the relationships and keep stakeholders as allies.

- **High Power/Low Ethics (HP/LE): Keep Satisfied, and Be Cautious (Prevent negative politics)**

  Stakeholders in this quadrant might cause problems because of their high power but low ethics. Therefore, project managers should keep them satisfied and pay special attention to them so they do not use their power to harm the project and relationships among stakeholders. These stakeholders are most likely to create negative politics to satisfy their interests.

5. **Power versus integrity matrix**

   Integrity is a powerful trait of leaders. It encompasses honesty, trustworthiness, and uprightness of character. People with high integrity walk their talk and choose to say and do the right thing, even if they have to compromise their interests. Generally, they have high levels of ethical standards, and they are sincere with others. Stakeholders with high integrity can help project managers succeed by providing them good mentorship and acting as a project champion.

   As shown in figure 5.5, when power levels are mapped with levels of integrity, stakeholders can be classified into four quadrants that show associated strategies to deal with stakeholders on the power/integrity matrix.

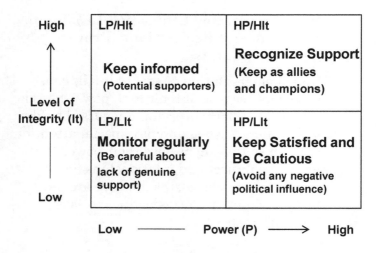

Figure 5.5. Matrix showing levels of power and integrity.

The following are associated strategies project managers can use to manage stakeholders:

- **Low Power/High Integrity (LP/HIt):**
  Keep Informed (Potential supporters)

  Stakeholders in this quadrant have low power but high integrity levels, and therefore, project managers should keep these stakeholders informed. Project managers should pay special attention to these stakeholders, use them as reliable resources in their project teams, and empower them, so when they are promoted in the organization and increase their power level, they can help the project managers on future projects. Furthermore, they might influence the people with low integrity to talk positively about the project and project manager.

196

- **Low Power/Low Integrity (LP/LIt): Monitor Regularly (Be careful about lack of genuine support)**

  Project managers should monitor stakeholders in this quadrant regularly because there is a risk that these stakeholders with low integrity might not support the project genuinely and might spread false rumors about the project, especially if they do not like the project manager. It is good that the stakeholders might be unable to directly influence project outcomes negatively because of their low power. However, project managers should be careful in dealing with these stakeholders.

- **High Power/High Integrity (HP/It): Recognize Support (Keep as allies/ champions)**

  These stakeholders are important, and project managers must recognize their support to keep them as allies and champions. They are sincere, genuine, and say and do the right things because of their high integrity levels, and because of their high power level, they can help the project manager succeed in meeting project objectives and in getting extra resources to meet special project challenges. Project managers should keep them informed about the project benefits and successful project management and maintain good working relationships to keep them as their allies and project champions.

- **High Power/Low Integrity (HP/LIt):
  Keep Satisfied, and Be Cautious (To
  avoid any negative political influence)**

  Project managers should be careful of
  stakeholders in this quadrant and keep them
  satisfied. The stakeholders might act like
  Sharks, abuse their power, and harm the
  project to satisfy their interests. They might
  create negative politics by spreading false
  rumors, especially if they do not like the
  project manager.

Project managers must pay attention to the
second category of attributes related to stakeholder
personal traits: predictability, commitment to
change, respect, ethics, and integrity. They should
prioritize stakeholders by mapping their level of
power and these attributes.

After analyzing stakeholders, project managers
must prioritize stakeholders by mapping their levels
of power and main attributes that can be divided
into categories. Category 1 contains attributes
related to stakeholder interests, and category 2
contains attributes related to stakeholder traits.

Chapter 5 Summary

Politics are inevitable in project environments
because most projects have stakeholders with different interests, expectations, and ways of doing things.
Project managers must understand the politics associated with successful management of stakeholders
because research has shown the following (Crawford
& Da Ros 2002, 20.1–20.10):[17]

5 - Politics in Managing Stakeholders

1. Organizational politics and acquisition of project resources are highly correlated.

2. Project managers' ability to manage organizational politics is crucial to project success.

   They should do the following:

   * Identify different categories of stakeholders:

     * Core

     * Internal

     * External

     * Unexpected

   * Analyze stakeholders for the following reasons:

     * To understand goals and interests

     * To analyze the political structure

   * Prioritize stakeholders using main attributes

     The attributes can be divided into the following two categories:

     Category 1: Attributes related to stakeholder interest:

       * Project knowledge

       * Interest

       * Influence

       * Impact

Category 2: Attributes related to stakeholder traits:

- Predictability
- Commitment to change
- Respect
- Ethical standards
- Integrity

Part II Summary

Most people view politics negatively. However, politics can be both positive and negative depending on how they are viewed and used. Negative politics focus on satisfying personal interests, whereas positive politics emphasize a culture of increased collaboration, teamwork, and synergy. The main theme of negative politics is to gain control over people, resources, information, and situations, whereas positive politics focus on the importance of effective communication and influencing to encourage teamwork and win-win solutions.

Establishing clear organizational strategies and goals, communicating them effectively, and gaining support and buy-in from project personnel are crucial elements of positive politics because they provide people direction and support. Senior management should create positive politics by using the Ten Commandments presented in Part II to create a team environment to achieve higher performance leading to a culture of increased collaboration, synergy, and cooperation.

Politics are inevitable in any organization, and they are caused by several factors related to four broad categories of issues:

1. Projects and project management issues
2. People behavioral issues
3. Organizational and management issues
4. External and global market environment issues

Because organizational politics cannot be eliminated, senior management should analyze and evaluate various issues that lead to politics, especially negative politics, and develop strategies to minimize the negative impact of politics. Project managers should develop good champions who are interested in the project and who can remove roadblocks as needed to achieve successful outcomes.

Politics are a necessary part of life in project environments because most projects have stakeholders with different interests, viewpoints, expectations, and personalities. Managing politics involves managing this diverse mix of stakeholders. Project managers must first identify different categories of stakeholders:

- Core
- Internal

- External

- Unexpected

They then must analyze the stakeholders in terms of their goals, interests, and expectations to determine how they fit in the political structure and who are true allies, movers, shakers, supporters, and barriers. To manage stakeholders successfully, project managers should prioritize stakeholders using the relationships between levels of power (low to high) and the following two categories of attributes:

A. Attributes related to project and stakeholder interests:

- Knowledge

- Interest

- Influence

- Impact

B. Attributes related to stakeholder traits:

- Predictability

- Commitment to change

- Respect

- Ethical standards

- Integrity

After prioritizing stakeholders, project managers should map stakeholder power levels with all attributes and develop appropriate communication strategies to manage them.

Part III. Understanding the Political Landscape & Stakeholder Behaviors

Chapter 6: Analyzing the Political Landscape

6.1. Three Political Positions

6.2. Who Engages in Negative Politics?

Chapter 6 Summary

Chapter 7: Understanding Political Behaviors

7.1. Political Behaviors at the Interpersonal Level

7.2. Political Behaviors Related to Team Environment

7.3. Political Behaviors Related to the Organization

Chapter 7 Summary

Chapter 8: Managing Political Behaviors

8.1. Initiatives at the Interpersonal and Team Levels

8.2. Initiatives at the Organizational Level

Chapter 8 Summary

Part III Summary

Understanding the Political Landscape & Stakeholder Behaviors

*To be persuasive we must be believable; to be
believable we must be credible; to be credible
we must be truthful.*
—Edward R. Morrow

Politics are part of life in project management
environments. Despite good planning,
scheduling, and budgeting, most projects fail
for political reasons, not technical reasons. Many
project managers dislike politics and consider
them damaging to the organization, but politics
are everywhere and important to manage projects
successfully in any organization.

Because project managers have great responsibility, but no direct authority over people working
on their projects, influence and politics are even
more crucial to manage projects effectively. Project
managers have no stable base of power, and so, they
must influence resource managers to get the right
resources in quantity and skill mix. Projects also
cross over many functional departments, and so,

project managers must negotiate and bargain for all resources.

Besides a formal organizational structure, every organization has its political landscape related to power structure and personalities at different organizational levels. Project managers must understand and analyze this political landscape and project stakeholders' viewpoints about politics. Often, projects have stakeholders with different interests, backgrounds, expectations, and personalities. These stakeholders behave differently, based on circumstances, as they manage their tasks and people. These behaviors might be perceived as political, leading to negative or positive politics.

People take three political positions, based on how they view politics, how they deal with politics, and how they manage their people and projects:

1. Flounders (politically naive)

2. Sharks

3. Dolphins (politically sensible)

Politically sensible managers must understand people who engage in negative politics and how they should deal with them.

Several political behaviors might lead to negative or positive politics. These behaviors are associated with how stakeholders do the following:

1. Protect themselves

2. Control and filter information

3. Push their ideas

4. Use their network power to meet their objectives, gain cooperation, and make decisions

Project managers must understand different stakeholders' political behaviors and the chief motives for those behaviors.
Political behaviors of stakeholders at different organizational levels can be classified in three categories:

1. Behaviors at the interpersonal level
2. Behaviors related to the team environment
3. Behaviors related to the organization

Practical techniques are described in this part to manage various political behaviors and deal with initiatives at the interpersonal and team levels and initiatives at organizational levels. Project manager should develop these techniques to meet organizational objectives with "the art of positive politics" and create a culture of collaboration, cooperation, and teamwork.

Analyzing the Political Landscape

*The roots of violence: Wealth without work,
Pleasure without conscience, Knowledge
without character, Commerce without
morality, Science without humanity, Worship
without sacrifice, Politics without principles.*
—Mahatma Gandhi

Politics are part of life in project environments because project team members have diverse backgrounds, expectations, and interests. Various political behaviors exist at all organizational levels. Most project managers are interested in planning and executing project activities to deliver results that should help meet organizational goals. They might not be great at playing organizational politics to get things done, but good project managers understand the importance and dynamics of organizational politics. They recognize that politics are often necessary to get things done. A big mistake people make is to assume politics don't exist and they need not learn any skills to understand and manage politics.

You must understand that politics can be both negative and positive, depending on how you approach them. Politics can negatively affect team productivity when people ignore them. *Flounders* (politically naive) dislike engaging in politics and believe that if they ignore them, they will disappear. They fail to anticipate the influence of politics on group dynamics and project outcomes. Alternatively, others who are self-centered use politics for personal gain. These people are *sharks*.

To succeed in any organization, you must understand the importance and dynamics of politics and develop skills to navigate politics and, thus, become politically sensible. *Dolphins* (politically sensible) can analyze the political landscape as to who has more power and how they use their powers and influence. They can anticipate the warning signals and position themselves in a politically charged work environment.

6.1. Three Political Positions

The possibilities are numerous once we decide to act and not react.
—Gloria Anzaldua

Various political behaviors affect people and projects differently. Negative behaviors create frustration, stress, wasted time, confusion, and anxiety, leading to a drop in everyone's overall performance, whereas positive behaviors encourage everyone to work together and create a culture of more collaboration and teamwork.

People act according to their political behaviors to meet objectives (personal, departmental, project, or organizational). Personality, viewpoint on politics, and approach to managing projects and solving organizational problems motivate political behavior. Verma (2014) and Pinto (1998, 75–77; 1994, 118–122) presented how people take three political positions, based on their viewpoints about politics and their approach to managing people and projects:[1]

1. Flounders

2. Sharks

3. Dolphins

Some project managers are politically naive (Flounders). These project managers think politics should be eliminated because politics are bad and wasted time. Because Flounders dislike politics, they ignore them and often encounter problems in managing their projects and people. In contrast, some people are Sharks. They use negative politics to meet their goals, even if at the cost of others and organizational goals. Dolphins are politically sensible and believe politics are part of life and necessary to get things done to meet project and overall organizational objectives. Typically, political behaviors that aim to resolve problems to help meet organizational goals are positive and acceptable.

Figure 6.1 shows a graphical representation often seen in an organizational political landscape— a pond in which Flounders, Sharks, and Dolphins swim. In this figure, Flounders lie near the ocean floor covered in mud. They represent politically

naive people who ignore and avoid politics and who
are passive about dealing with politics. Dolphins are
quick and represent politically sensible managers
who believe politics are essential to organizational
life, and it is important to develop skills in navigating through politics to manage projects.

Figure 6.1. Swimming in a political pond.

In figure 6.1, Sharks swim in the middle and try
to harm Flounders and Dolphins. In this context,
Sharks represent self-centered people/managers who
want to get ahead at any cost. They do not hesitate
to attack Flounders and Dolphins to meet their
objectives. People with Shark behavior try to manipulate people and information to their advantage.
They find Flounders easier prey than Dolphins because of their passive nature and because Dolphins
are nimble, and they might even harm Sharks by
working as a team.

Dolphins are agile and use their speed and skills to protect themselves from Sharks. Similarly, politically sensible people try to use politics to further organizational goals as the highest priority and then think of department goals and project goals. They continually expose Sharks whenever they try to manipulate or twist the information or situation to steal critical resources from others. Politically sensible project managers are great in aligning their project goals to organizational strategies and try to get protection from their champions well respected in the organization with more positional power than Sharks have.

6.1.1. Flounders (Politically Naives)

Politics is perhaps the only profession for which no preparation is thought necessary.
—Robert Louis Stevenson

Politically naive people are like flounders, a species of ocean-dwelling fish. Flounders use the passive technique of camouflage to deal with predators, covering themselves in mud and lying on the ocean floor (Wikipedia).[2] Similarly, politically naives are passive in dealing with politics. They either ignore politics or avoid them. They are not careful to protect themselves from political tactics used by others in power or those who might use manipulative strategies, and so, they are more likely to be political victims.

Pinto (1998, 75–77; 1994, 118–122) and Verma (2014) described how Flounders' view politics, how they deal with politics, and how they manage their projects and people.[3] These viewpoints and approaches are summarized in table 6.1.

Viewpoints on Politics	How They Deal with Politics	Main Approach in Managing Their People and Projects
Politics • are bad/negative • are wasted time • are unpleasant • are undesirable	• Ignore and avoid • Hope they will disappear • Condemn them	• See themselves as direct and straight shooters • Tell everything as it is • Trust their stakeholders too readily • Over depend on processes, tools and methodologies • Analyze and justify their projects only in tangible costs/benefits

Table 6.1. Political Viewpoints and Management Techniques of Flounders (Naives)

1. How Flounders view politics

   Politically naive people don't like politics. Flounders believe organizational politics are bad, and they should be eliminated. They believe politics are slow and undesirable torture in which dishonesty, selfishness, secrecy, and deception replace logic, discipline, transparency, and loyalty (Myers 2012).[4] Their viewpoints about politics, while working with stakeholders, can be summed as follows:

   • Politics are bad because they create problems.

- Politics are wasted time, and they lead to frustration and stress.

- Politics are unpleasant because they create conflicts and bad feelings.

- Politics are negative.

- Politics are undesirable, and they should be eliminated or not allowed in any organization.

2. How Flounders deal with politics

   Politically naive people do not like politics because they believe they are a game that wastes much time and energy. They would like politics to disappear. They deal with politics in the following ways:

   - Ignoring and avoiding them

   - Hoping they will disappear

   - Condemning them

   Because Flounders dislike politics, they do not want to understand their dynamics and learn to deal with them. Instead, they want politics eliminated and disallowed. However, politics are part of corporate life. They have been around for ages, and they will continue to exist. Naive project managers should recognize that politics will never disappear, and they should learn to deal with them before they disappear themselves or become a political victim.

3. How Flounders manage their people and projects

   Naive project managers prepare the project plan with good estimates and risk analysis to the best of their knowledge and conclude on their own that their project plan is good and their project is good for the organization. In addition, they believe they can manage the project, and then, they expect the organization to support them. Naive project managers manage their people and projects in the following ways:

   - They see themselves as direct and straight shooters.

   - They tell everything as it is without evaluating the indirect impact.

   - They trust their stakeholders too readily without carefully evaluating the depth of relationships.

   - They over depend on processes, tools, and methodologies.

   - They analyze and justify their projects only in tangible costs/benefits rather than understanding the long-term success criteria.

   Flounders are honest people, but they do not recognize that despite good planning, a straightforward communication style, and good project management tools and processes, politics derail many projects. They fail to see the political signals and learn their lessons only after problems and failures. They cannot survive in politically charged environments unless they change their outlook about politics. To avoid becoming a political victim, they should first realize that politics are part of life in project environments and then recognize that they are necessary to do things through others to meet project objectives.

6.1.2. Sharks

People who hurt others have themselves been hurt. People who do not love themselves cannot show love to others. And people who do not have any self-respect have no idea how to give respect to others.

—Robin Sharma

Sharks are some of the world's most infamous predators. There are more than 500 species of sharks; many hunt alone, though some work in packs. Just as sharks are ultrasensitive to the smell of blood, political Sharks in corporate environments pay special attention to opportunities to advance themselves, even at the cost of organizational goals (Wikipedia).[5]

Pinto (1998, 75–77; 1994, 118–122) and Verma (2014) described how Sharks view politics, how they deal with politics, and how they manage their project and people.[6] These viewpoints and approaches are summarized in table 6.2.

Viewpoints on Politics	How They Deal with Politics	Main Approach in Managing Their People and Projects
• Politics can be used for self-serving purposes • Politics are OK to use for self gains • The aim is to win in political games	• Use politics to advance their interests and agenda • Use politics to disassociate themselves when projects are in trouble • Use politics to deflect account-ability issues • Use politics to self promote • Use politics to get information and opinions from others	• Manipulate information and people • Twist information and ideas to their advantage • Put a different spin on information • Take credit for others' good ideas • Feel OK stealing ideas and plagiarizing • Flatter power brokers at senior management levels • Talk badly about others in their absence • Disassociate themselves from project failures or problems • Blame others and point fingers • Deflect accountability issues • Extract useful informa-tion and opinions from others • Withhold information

Table 6.2. Political Viewpoints and Management Techniques of Sharks

1. Shark viewpoints on politics

   Sharks like politics, and they consider politics an opportunity to advance themselves, even if at the cost of organizational goals and strategies. They are predatory, and they don't care if their shark behavior and actions harm

others. Their viewpoints about politics can be
summarized as follows:

- Politics can be used for self-serving
  purposes.

- It is OK to use politics for self-gains, even
  at the cost of others.

- The aim is to win in political games, just
  as in love or war.

2. How Sharks deal with politics

   Sharks are predatory and selfish. They use
   politics to meet their hidden agenda and
   self-interests. People with politically Shark
   behavior deal with politics as follows:

   - Use politics to advance their interests and
     agenda

   - Use politics to dissociate themselves when
     projects are in trouble

   - Use politics to deflect accountability
     issues

   - Use politics for self-promotion

   - Use politics to get information and opin-
     ions from others for their advantage

3. How Sharks manage their people and projects

   Sharks are self-centered. They create wrong
   perceptions to make themselves look good.
   People with Shark behavior use these
   techniques to manage their people and

projects in the following ways (Verma 2014; Myers 2012):[7]

- They manipulate information and people.
- They twist information and ideas to their advantage.
- They put a different spin on information to give a perception to others that they are more interested in true teamwork in meeting organizational goals.
- They take credit for others' good ideas.
- They feel OK stealing ideas and plagiarizing.
- They flatter power brokers at senior management levels to meet their self-interests.
- They talk badly about others in their absence.
- They dissociate themselves from project failures or problems.
- They blame others and point fingers when problems happen.
- They deflect accountability issues.
- They extract useful information and opinion from others to use to their advantage.
- They withhold information to increase their power.

Sharks follow and create negative politics. Their priority is to win and make themselves look good even at others' cost. Shark behavior can demotivate people and jeopardize meeting organizational goals.

Just as project managers must analyze the political landscape to manage politics, sales managers should analyze their market landscape to analyze their competitors and clients. Salespeople should know their competitors' strategies and plans and evaluate their clients' needs continuously to increase their sales. Like project managers competing for the same limited human resources in an organization, salespeople compete for the same market share and clients.

Harvey McKay (2005, ix–xviii), in his book *Swim with the Sharks Without Being Eaten Alive*, compares competitors and major customers with sharks in that each is self-centered and looking for best terms and deals to meet their objectives, even at others' cost. Dealing with Sharks requires good understanding of two factors:[8]

1. Changes in the business environment — There have been many changes in the past 15 to 20 years in the business world (McKay, 2005, ix–xviii):[9]

    *   Millions of people have lost jobs because of mergers, acquisitions, joint ventures, restructuring, downsizing, and so-called *right sizing*.

    *   Millions of jobs have been lost because of outsourcing manufacturing and simple operations to foreign countries such as India, China, Malaysia, and many other countries in South America, Eastern Europe, and Asia.

    *   According to labor statistics, today's graduates face 10.3 job changes in their

lifetimes, not including up to 5 career changes.

- Most organizations have a multigenerational workforce with a different outlook about work-life balance, personal goals, and values. They prefer different communication strategies and tools. Project managers and sales managers face special challenges to communicate with and motivate a multigenerational workforce.

- Managers cannot depend on secretaries for typing and simple office work. Instead, they are expected to do a large share of these jobs, and computer literacy is a required basic skill.

Besides these changes, there have been many changes in technology, marketing, human resource management, project management, and globalization. Those unprepared to compete in the business world are more vulnerable than ever to predators eager to take away business and jobs from others. Similarly, project managers should know major changes in their business environment and evaluate the impact on their projects. They should learn to become politically sensible and develop good conceptual and people skills to prepare themselves well to meet these challenges and avoid takeover by Sharks.

2. Evolution of Sharks — The evolution of Sharks stopped a few hundred years ago. To deal with Sharks, managers must figure out the new Shark skills. The good news is that

the basic Shark behavior and skills are the same in that Sharks are selfish and predatory. Some managers in project environments act like sharks and use organizational resources and systems to their advantage, even at the cost of other projects and sometimes the organization.

In sales, executives work to increase their market share and client base at others' cost. They often do not hesitate to steal others' market share in the world of limited market demand because it is a zero sum game, that is, if others gain a certain segment of the market share, they get no part of it. Similarly, customers have many choices, and they look for the best deals, only choosing the supplier who offers them the most suitable options in price and quality for the product and services they want. Some customers show loyalty to some suppliers, but then they expect something major.

However, creative salespeople devise strategies to create additional demand for their products and services rather than stealing from others, leading to a win-win situation. Similarly, politically sensible project managers increase the capacity of organizational resources by cross training and mentoring and then use politics to further the organizational strategies and goals, rather than using limited resources to push their projects.

Harvey McKay suggested invaluable ideas to survive and thrive in today's competitive global environment. He described sixty-nine lessons with interesting stories covering salesmanship, negotiation, and general management. In addition, he suggested nineteen quick life lessons (quickies) and excellent ideas to share with kids—"Help your kids beat the odds," and prepare them to meet challenges in life and the workplace.

6.1.3. Dolphins (Politically Sensibles)

> *Satisfaction lies in the effort, not in the*
> *attainment. Full effort is full victory.*
> —Mahatma Gandhi

Dolphins are clever animals whose greatest skill is effective teamwork. Rather than fending for themselves individually, dolphins use a collaborative approach to hunting to maximize results (Wikipedia).[10] Politically sensible managers, like Dolphins, are fast, intelligent, and careful. They strive to meet or exceed stakeholder expectations. They focus on working together and creating synergy to meet organizational goals. Verma (2014) and Pinto (1998, 75–77; 1994, 118–122)) described how Dolphins view politics, how they deal with politics, and how they manage their projects and people.[11] These viewpoints and approaches are summarized in table 6.3.

1. How Dolphins view politics

    Politically sensible managers recognize that politics are a necessary part of life in corporate environments. They accept that politics exist; therefore, it is better to understand the dynamics of politics and learn to manage politics to do things to meet project objectives and organizational

Viewpoints on Politics	How they Deal with Politics	Main Approach in Managing their People and Projects
• Politics are a natural way to deal with corporate work • Politics can be positive and helpful in managing projects • Politics are necessary to get things done by others with different interests • Politics can create a culture of collaboration and teamwork	• Use politics to strengthen network • Use politics to gain support from people with power and influence • Use politics to increase profile of project and team • Use politics to further organizational goals (See figure 6.2 for the sequence of priorities/steps when dealing with politics.)	• Understand and analyze the political landscape • Strengthen network and working relationships • Recognize the importance of informal powers • Master the art of influencing to gain support and cooperation • Think and look for win-win solutions • Develop good communication skills • Consider conflict natural, and manage it • Prepare for the worst • Keep friends close and enemies closer • Respect project stakeholders and do not alienate them

Table 6.3. Political Viewpoints and Management Techniques of Dolphins (Politically Sensibles)

objectives. Their viewpoints about politics can be summed as follows:

- Politics are a natural way to deal with corporate work.

- Politics are not bad; they can be positive and helpful in managing projects.

- Politics are necessary to get things done by others with different interests.

- Positive politics can create a culture of collaboration and teamwork.

2. How Dolphins deal with politics

   Politically sensible people use politics to further organizational goals as the highest priority, and then department goals. Politically sensible managers use politics to help themselves and others by working hard to achieve organizational goals. They deal with politics as follows:

   • They use politics to strengthen their network.

   • They use politics to gain support of people with significant power and influence.

   • They use politics to increase the profile of their project and their team.

   • They use politics to further the organizational goals in this sequence:

     1. Overall organizational goals and strategies

     2. Department goals and strategies

     3. Project goals

     4. Personal goals

   Figure 6.2 shows the sequence of priorities Dolphins use when dealing with politics. Dolphins emphasize organizational strategies and goals and then departmental goals, because after the project finishes, they perhaps must go back to their department for the next assignment. Flounders keep talking about their project only, and Sharks focus on their goals and objectives.

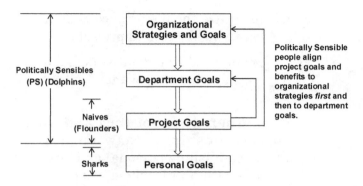

Figure 6.2. Politically Sensibles' priority sequence for using politics.

3. How Dolphins manage their people and projects

    Politically sensible managers keep their eyes and ears open to look for Sharks ready to play negative politics to meet their objectives. They increase their circle of influence and create good working relationships with everyone in their network. Verma (2014), Myers (2012), and Pinto (1998, 146–152) suggested techniques that politically sensible managers should use to manage politics. Politically sensible managers use these strategies to manage their people and their projects:[12]

    • **Understand and analyze the political landscape** — Project managers should analyze their organization's political land-scape and develop appropriate political tactics to manage their projects.

    • Stop being politically naive and recog-nize that politics are inevitable.

- Recognize the importance of politics in delivering successful projects.
- Understand well what stakeholders say and what they want.

- **Strengthen their network and working relationships** — Networking is about establishing personal and business contacts with people whom you can call on to ask for favors to deliver successful projects. It is about giving and receiving, so first, give or do things for others. Only then, can you receive something from others.

  Politically sensible managers recognize the power of networking and good working relationships. They focus on developing strong networks with people in power and with a large circle of influence. To do this, they nurture and feed their network continually. At the project level, they do favors for other stakeholders who are most likely to reciprocate to get information, resources, and expert opinions.

- **Recognize the importance of informal powers** — Project managers should develop their informal power and help their team members increase their informal powers. They also use their power to gain support from stakeholders long term, that is, they mostly use their power informally, but they are not shy to use it formally, only as a last resort.

- **Master the art of influencing to gain support and cooperation** — Project managers should learn the art of influencing their stakeholders to gain their support and commitment long term. To do this well, they understand that people will usually do whatever is in their best interest. Everyone looks for What's in It for Me (WIIFM). Therefore, to influence their stakeholders, project managers should discover What's in It for Them (WIIFT) and the reason they should do what you want them to do, and then take appropriate steps. Project managers can influence their managers, team members, and other stakeholders by using these five methods suggested by Keys and Case (1990, 38–51):[13]

1. Establish a reputation as an expert.

2. Prioritize and strengthen relationships with influential people in the organization who can help meet their goals.

3. Network with experts and resource managers.

4. Understand the influencing process and use the appropriate influencing strategy according to the people and situation.

5. Know their people well. Be socially sensitive, articulate, and flexible.

- **Think and look for win-win solutions**
  — The win-lose approach creates enemies who look for opportunities to undermine you. Therefore, either destroy them fully or use a win-win approach to create friends. In developing negotiating strategies, project managers should emphasize fairness for both parties and use mutually acceptable win-win outcomes. Fisher and Ury (1981) suggested using principled negotiation to reach win-win outcomes:[14]

  - Focus on interests, and not on positions.

  - Separate the people from the problem.

  - Generate options that advance shared interests.

  - Use objective criteria, and be fair.

  - Think of long-term relationships rather than taking advantage to meet personal goals.

- **Develop good communication skills**
  — Project managers must recognize that communicating effectively and, especially, listening are critical to gain support and commitment from their project stakeholders. To do this well, they use these guidelines:

  - Involve their stakeholders in all phases of the project life cycle (PLC).

  - Ensure that mutual expectations are clear to one another.

- Make stakeholders feel free to give suggestions and feedback to resolve project problems and issues.

- Follow core values and principles to evaluate people accordingly.

- Create high-performance teams.

- **Consider conflict natural, and manage it** — According to Pinto and Kharbanda (1995), Project managers should understand the dynamics of conflict and recognize it as natural and healthy.[15] Project teams are composed of members with different interests, personalities, and backgrounds. They might come from different functional departments accustomed to doing things differently. In such cases, conflicts are natural.

  Many project managers view conflict as signs of disintegration and project failure and, therefore, respond to conflict with panic and stress. They should not suppress the conflict, hoping it will go away. Instead, project managers should understand the cause of conflict and the personalities involved. They should understand the two dimensions of conflict management:

  1. Time (long-term or short-term results)

  2. Standard of fairness (being fair and unfair)

There are four conflict management styles, based on these two dimensions, as shown in figure 6.3. Politically sensible project managers use an appropriate conflict management style to balance between the time (long-term and short-term results) and standard of fairness (being fair or unfair).

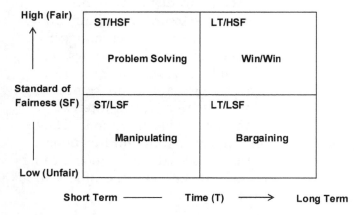

Figure 6.3. Conflict management styles based on time and standards of fairness.

- **Prepare for the worst** — Expect and prepare for surprises in relationships, and learn to let go. Know some follow negative politics to get ahead at any cost.

- **Keep friends close and enemies closer** — Sun Tzu, the author of *The Art of War* (2002), illustrated this concept well, suggesting that politically sensible managers should look out for Sharks, understand their motives, and then develop an appropriate plan to deal with them.[16]

- **Respect project stakeholders, and do not alienate them** — Project managers can earn their stakeholders' respect and keep them as allies by using the following approach:

  - Don't say bad things about others because it backfires.

  - Be empathetic to sensitive personal issues.

  - Stay neutral when dealing with poor performers by using these steps:

    - Use "I" statements, instead of "you."

    - Don't rehash the past.

    - Don't editorialize or use judgmental statements.

    - Be silent, and wait for a response.

Dolphins continually develop and exercise their informal power. However, they use their formal power, if needed, only as a last resort. Dolphins also continually expand their network by nurturing and maintaining relationships, though they know their network's quality is more important than its size is. Dolphins and their teams have mutual trust because they understand that working together as a team is important. Similarly, Dolphins recognize their team is their power base. Therefore, they focus on developing high-performance teams by providing proper support and challenging opportunities, demonstrating trust and confidence, and nurturing

creativity and innovation. Politically sensible people
are like dolphins because they are flexible and can
quickly adapt to changing circumstances and their
team members' maturity level.

Three political positions are often encountered
in project management environments. Politically
naive people (Flounders) like to pretend politics do
not exist. Sharks manipulate people and processes
to their advantage. Politically sensible managers
(Dolphins) are fast thinkers and great at navigating
organizational politics.

6.2. Who Engages in Negative Politics?

We only confess our little faults to persuade
people that we have no big ones.
—Francois De La Rochefoucauld

Several political behaviors are commonly found
in many organizations. Organizational politics
flourish under the right conditions (Hardy 1990, 14;
Goh & Doucet 1986, 77–86; Jick & Murray 1982,
141–169; Gandz & Murray 1980, 237–251).[17] People
continue to use political tactics if organizations toler-
ate or encourage them (Ferris, Russ & Fandt 1989,
143–170; Mintzberg 1985, 133–154).[18] Therefore,
senior management must discourage negative politics. The personality and personal values of project
stakeholders can influence their likelihood of engaging in organizational politics (House 1988, 305–357;
Porter, Allen & Angle 1981, 120–122).[19] Some people
usually get involved in negative politics because they
see politics as one of the few ways to get ahead in the
organization. Table 6.4 shows the characteristics of
people engaged in negative politics.

Types of People	What They Do
People who are power-hungry	Use unfair means to gain more power
People with internal focus of control	Demonstrate over-confidence and control over situations
People who do not share information	Hoard information to show their power
People with Machiavellian values	Deceive and manipulate others for personal benefits
Unsatisfied and unhappy employees	Tend to spread wrong information and rumors

Table 6.4. Characteristics of People Engaged in Negative Politics

1. **Power-hungry people** — Having power is not bad. Project managers should increase their power, especially informal or personal powers. They should exercise power appropriately, that is, mostly informally, to maintain it and earn respect from project stakeholders. So, having more power and knowing how to use it helps a great deal in getting projects done with others over whom project managers have little or no direct authority.

   All project managers should be power-hungry, but the ones who get involved in negative organizational politics are those who try to steal power from others or gain power through unfair means. For example, regarding expert power, they steal the credit from others to create the perception they are the experts. Project managers with this characteristic might enjoy some short-term benefits until other stakeholders learn the truth and change their perceptions about them.

2. **People with internal focus of control —**
People with internal focus of control believe
they are in charge of their destiny, and they
are determined to accomplish their goals by
hook or by crook. The "by crook" part urges
them to get involved in negative organizational politics. They are overconfident and
control situations. These characteristics
should be understood, as people should be
strongly determined to meet their goals and
not leave everything to fate. People should
not use unfair means to meet personal objectives, especially at the cost of organizational
objectives.

3. **People who don't share information —**
Information is power. Some people like to
hide or hoard information to show off their
power. The chief reason for this behavior
is that they typically feel insecure about
their position in the organization. They are
usually unwilling to share information readily because they believe they can leverage
their position of having information to their
advantage. Information is an informal power;
the more information you share, the more you
get, leading to increased information power
and the overall power package.

4. **People with Machiavellian values —**
Machiavellianism is named after Niccolo
Machiavelli, the sixteenth-century Italian
philosopher who wrote a well-known treatise
on political behavior, *The Prince*. People
with Machiavellian values seldom trust
their coworkers and manipulate people and

information to meet their goals, even at the cost of organizational goals (Mudrack 1993, 517–524; Christie & Gels 1970).[20] They believe lying and deceiving are acceptable, and nothing is wrong with using these techniques and manipulating people for personal gains. Such people indulge in negative organizational politics because they see them as an easy way to meet their objectives.

5. **Unsatisfied and unhappy employees** — Unsatisfied employees engage in negative politics because they do not agree with the management, overall organizational structure, policies, and procedures. The following are major disagreements:

   • Disagreements related to senior management styles

   • Disagreements related to organizational strategies and long-term plans

   • Disagreements related to selection of programs and projects

   • Disagreements related to priorities

   • Disagreements related to administrative policies and procedures

   • These unhappy people create negative perceptions about management, leading to overall dissatisfaction and negative energy.

Project personnel get involved in negative organizational politics for many reasons, depending on their nature, desire to meet personal objectives in the organization, and the emphasis of maintaining their powers for the short term or long term. Organizational politics flourish under these conditions. People who often engage in negative politics cannot maintain their power and lose their project stakeholders' respect and trust, further reducing their abilities to influence project stakeholders long term and to do things in project environments. Senior management should lead by example by clarifying goals, roles, and responsibilities to avoid confusion, misunderstanding, anxiety, and stress.

Chapter 6 Summary

People take three political positions, based on their viewpoints about politics and their approach to managing people and projects:

1. Flounders (politically naïve)

2. Sharks

3. Dolphins (politically sensible)

Flounders ignore and avoid politics, and they are passive in dealing with politics. Sharks are self-centered people who want to get ahead at any cost, even by harming Flounders and Dolphins. Dolphins are politically sensible managers who believe politics are essential to organizational life and you should develop skills in navigating through politics to manage projects. Each political type is distinguished by how it views politics, how it deals with politics, and how it manages its people and projects.

According to Harvey McKay,[21] you must understand two factors to deal with Sharks:

1. Changes in the business environment

2. Evolution of Sharks

Dolphins use the following strategies suggested by Dale Myers and Jeff Pinto[22] to manage politics and their people and projects:

1. Understand and analyze the political landscape.
2. Strengthen their network and working relationships.
3. Recognize the importance of informal powers.
4. Master the art of influencing to gain support and cooperation.
5. Think and look for win-win solutions.
6. Develop good communication skills.
7. Consider conflict natural, and manage it.
8. Conflict management has two dimensions:
9. Time (long-term or short-term results)
10. Standard of fairness (being fair and unfair)
11. Prepare for the worst.
12. Keep friends close and enemies closer.
13. Respect project stakeholders, and do not alienate them.

We identify five types of people who engage in negative politics:

1. Power-hungry people.
2. People with internal focus of control.
3. People who don't share information.
4. People with Machiavellian values.
5. Unsatisfied and unhappy employees.

Project personnel get involved in negative organizational politics for many reasons. Senior management should lead by example by clarifying goals, roles, and responsibilities to avoid confusion, misunderstanding, anxiety, and stress. They should create a culture of collaboration, synergy, and teamwork to rechannel project personnel energy from the negative side to the positive side to focus on delivering successful results.

Understanding Political Behaviors

How, when, and where you say
something can actually be more
important than the message itself.
—Anne Bruce and James S. Pepitone

In a business environment, change is natural and sometimes necessary to thrive and survive. It forces us to change our goals and plans. Sometimes, political behaviors of senior management and high-profile stakeholders influence changes that might be good for the organization. Different changes highly influence organizations and projects (Doyle 1990, 36–45; Offerman & Gowing 1990, 95–108).[1] Four primary factors—changes in goals and objectivity, staffing levels, budgets, and schedules—and secondary factors such as quality, risk, and contract management often influence project plans (Bakers 1992, 211–220; Bennis, Benne & Chin 1961, 69).[2] Change can be motivated by organizational politics and can be managed by influencing people's political behavior.

Many organizational behavior (OB) researchers have identified various typical political behaviors project stakeholders and others display in most organizations. Political behaviors can have a negative or positive impact on project performance and lead to negative politics, positive politics, or both. Therefore, these behaviors must be identified and then managed as soon as possible.

These behaviors can be grouped in the following three categories (Verma 2014; McShane, Steen & Tasa 2014, McShane 1995; Murray & Gandz 1980, 11–23; Allen, Madison, Porter, Renwick & Mayes 1979, 77–88):[3]

1. Political Behaviors at the Interpersonal Level

2. Political Behaviors related to the Team Environment

3. Political Behaviors related to the Organization

7.1. Political Behaviors at the Interpersonal Level

*Power is not revealed by striking hard
or often but by striking true.*
—Honore de Balzac

People show certain political behaviors when they deal with others and projects at the interpersonal level in difficult situations. People might behave in such a way depending on their experiences, working relationships with others, and their personalities and habits. Project managers should take time to understand their stakeholders and see things from their perspective to gain their support and minimize the impact of negative behaviors. Various political behaviors and their impact at the interpersonal level are shown in table 7.1.

Behaviors at the Interpersonal Level	Impact
Finger-Pointing (attacking or blaming others)	Negative
Name-Dropping	Negative/Positive
Introducing Change	Negative/Positive
Networking (establishing contacts) to Help Organizations	Negative/Positive
Sharing Information	Negative/Positive

Table 7.1. Political Behaviors and Impact at the Interpersonal Level

7.1.1. Finger-Pointing (attacking or blaming others)—negative impact

> *If you always blame others for your mistakes, you will never improve.*
>
> —Joy Gumz

Attacking, blaming, or bad-mouthing one another leads to negative politics because it makes people defensive, and they might reciprocate. It does not address or resolve issues constructively. Instead, it deflects the issues and confuses account-ability. When projects are not going in the right direction, political players or those responsible for the mistakes should hold themselves accountable. Instead, they transfer the blame to others and find scapegoats to defend themselves or use arguments to find a different cause for the problem (Ashforth & Lee 1990, 621–648).[4]

7.1.2. Name-Dropping—negative or positive impact

> *Power is not revealed by striking hard or often, but by striking true.*
> —Honore de Balzac

Some project managers face difficulties in getting support from project stakeholders and project personnel. They might not influence stakeholders to have them complete work. They name-drop, mentioning the senior manager names when requesting work be done, especially if these employees worked under them.

The name-dropping technique can have positive and negative impacts. For example, it works positively if the project manager gets the senior manager approval to use their names in seeking help from their subordinates. However, it can backfire on the project managers and create negative politics if stakeholders learn that the project manager never discussed it with those senior managers and did not get their approval to ask the subordinates directly to do tasks.

This behavior can also create a lack of trust between the project managers and project stakeholders and lead to wrong perceptions. Therefore, project managers must evaluate the political impact of using senior manager names and get their approval before asking their subordinates to do the tasks the project manager wants done.

7.1.3. Introducing change—negative or positive impact

> *It is always easier to talk about*
> *change than to make it.*
> —Alvin Toffler

Change is typical in most project environments. Change can be related to project processes, methodologies, objectives, and resources. Some project managers believe they must introduce changes to meet project objectives. However, they must recognize the potential resistance to change and the related stress and anxiety. Introducing change can have both a positive and a negative impact. Project managers might try to enforce this change to push their ideas, and then, encounter resistance. These barriers might be related to the following:

- Lack of understanding about reasons to resist changes

- Lack of preparation in developing the change process

- Lack of open communication and dialogue

- Lack of participation and buy-in

Project managers must recognize that resistance to change is inevitable. They must evaluate the situation and think about strategies to overcome barriers to change. The following are practical tactics to reduce the resistance to change:

- Explain clearly reasons behind the proposed changes. (Be honest!)

245

- Explain positive aspects of change for people and project performance.
- Openly address questions, concerns, and frustrations.
- Involve people and seek genuine feedback to develop and implement the change process.
- Do not use fear, threat, or coercion in pushing personal agendas.
- Use leadership skills to create a common and shared vision and then convince people.

If changes are introduced without explaining the reasons and benefits, stakeholders do not accept or cooperate in implementing the changes. They perceive efforts to push ahead with changes under such conditions as negative politics. If managers introduce changes by involving stakeholders and communicating with them about the potential benefits, stakeholders accept the changes and cooperate in the implementation. This technique is perceived as collaborative, productive, and positive.

7.1.4. Networking (establishing contacts) to help organizations—negative or positive impact

> *A friendship can weather most things and thrive in thin soil—but it needs a little mulch of letters and phone calls and small silly presents so often—just to save it from drying out completely.*
>
> —Pam Brown

Networking in a project environment can be internal and external and can be perceived as positive or negative. Networking leads to positive outcomes unless done with malicious intentions. It is amazing how much networking can help project personnel manage their projects, work packages, and tasks.

If people use their network to satisfy self-interests rather than the interests of the project and the organization, the behavior leads to negative politics. However, networking is often considered a positive trait of project managers.

Although project teams have members with different skills and expertise, there might be a need for special skill and information. By having trusting networks, people can receive valuable information to increase their expert power (Krackhardt & Hanson 1993, 104–111; Kaplan 1984, 37–52).[5] Project managers can contact people with different skills in their network to get help as needed to resolve difficult project issues and problems.

Three elements of networking in project environments can improve overall performance and produce better results:

1. Do and receive favors in supplying information, ideas, and tips.

2. Establish personal and business contacts to call on for help.

3. Share many skills, much knowledge, and experience.

"Do and receive favors" is the most important element of networking, meaning that project managers must first do favors for people in their network, and only then, can they expect to receive help. Often, it might not be the people project managers helped directly. However, if project managers are willing to help others often when approached, they can also find someone in their network to help them.

Networking is an important form of power. Project managers must recognize the importance of developing a good network of people from different industries with various skills and expertise. It helps them solve many project problems. They should follow two chief principles of effective networking:

1. **Nurture the network** — *Nurturing* refers to keeping the network alive, so project managers must stay in touch with their network. To do it more efficiently, they might need to categorize several people in the network and make it a priority to regularly communicate with them personally and professionally.

2. **Feed the network** — *Feeding the network* refers to doing favors for others and helping them as needed and approached to receive help when needed. First, make deposits in the networking account so your checks do not bounce.

Good project managers recognize that networking is an important power that can help them meet their project objectives. They understand the three important elements of networking and follow two main principles for developing and strengthening their network power.

7.1.5. Sharing information (negative or positive)

It is wise to disclose what cannot be concealed.
—Johann Friedrich Von Schiller

Information is power and how it is used makes a big difference on project outcomes. Different stakeholders use their information power differently, leading to negative or positive perceptions about their behavior. Some stakeholders believe withholding information is good for them and increases their power because others must depend on them to get that information. Mostly, this belief stems from feelings of insecurity and lack of self-confidence.

This behavior is negative because other stakeholders would also hide or withhold their information, reducing the overall knowledge pool to do things. However, sometimes, it is acceptable not to share classified or confidential information for security and other business reasons.

Other stakeholders believe information is an important informal power and the more they share, the more they get. These stakeholders share information with their team members, management, and other stakeholders more readily. This behavior is positive and leads to positive politics because others will also share their information, increasing cooperation and teamwork. Sharing information also increases the knowledge of both parties, which increases their expert power to manage projects.

Senior management must understand and analyze these political behaviors and create a climate of positive politics by encouraging positive behaviors and discouraging negative behaviors. They should establish organizational policies to reward senior managers who practice positive behaviors and influence other managers to do the same.

People behave differently at the interpersonal level to manage uncertainties, difficult situations, and politics in project environments. Various political behaviors at the interpersonal level have a positive or negative impact on project performance. People should be sensitive to others' feelings, and they should treat others the way they want to be treated. Project managers should know their project personnel's behaviors at the interpersonal level, understand their motives, and encourage positive behaviors to minimize the impact of negative behaviors.

7.2. Political Behaviors Related to Team Environment

Achievement is a WE thing, not a ME thing, always the product of many heads and hands.
—J. W. Atkinson

Projects are done by teams with diverse personalities, work habits, and levels of professional competence. People behave differently when they interact with others to carry out project activities in a team environment, which affects overall team performance. Such activities might involve gathering and disseminating information and acting to improve things. Project managers must understand their team members well and analyze their behaviors. They should try to encourage positive behaviors and minimize the impacts of negative behavior. Various political behaviors at the team level and their impact are shown in table 7.2.

Behaviors in a Team Environment	Impact
Controlling Information	Negative/Positive
Filtering Information	Negative/Positive
Incremental Improvements	Negative/Positive
Forming Coalitions	Negative/Positive
Churning (organizational amnesia)	Negative

Table 7.2. Political Behaviors and Impact Related to a Team Environment

7.2.1. Controlling information—negative or positive impact

> *Communication works for those*
> *who work at it.*
> —John Powell

Some project managers are control freaks. They want team members to send all information to them first for review, and then they decide what and how much information to send to whom and in what format. They do not want team members to send information to anyone on their own because some data is sensitive, and sending it to more people might affect project performance. For example, some information threatens their power, or it is damaging. If managers like to avoid deciding on a particular point, they put that issue or topic at the bottom of the agenda so the committee either gets tired or runs out of time.

Controlling information can lead to both negative and positive politics, depending on the situation. For example, it leads to a negative impact if the project manager is not fast enough to review the information and then disseminate it to required stakeholders promptly, as they might need that information to complete their project assignments to meet project objectives. The behavior is positive if the project manager does not become a roadblock in the flow of required information and is fast enough to review and send to everyone as needed. Here, they minimize the uncontrolled flow of information that might create politics and reduce project performance.

7.2.2. Filtering information—negative or positive impact

How well we communicate is determined
not by how well we say things but how
well we are understood.
—Andrew Grove

Filtering information commonly causes negative and positive politics. Sometimes, project managers use a strategy of overloading their sponsors or steering committee members with too much data so they depend on project managers to sort it and help them understand it. Some project managers or team leaders filter the information they receive from their team members or other stakeholders. They decide not to disseminate all the information in the same format and details they receive and edit it before sending it to other stakeholders.

Some project managers filter information because some sensitive information affects the project's overall schedule and budget. They decide in their wisdom that disclosing that information is not a good idea. They want more time to analyze and evaluate the information before sending it to others. Such behavior can lead to negative politics if that information goes to people who do not favor the particular project and want the project manager to fail.

Negative politics can be avoided if the project manager evaluates the information and mitigates the risks associated with it before letting it go to the stakeholders. The decision to filter the information and not tell everyone before evaluating it thoroughly can be positive, because the project manager could try to correct the situation and not create unnecessary anxiety among people.

7.2.3. Incremental improvements—negative or positive impact

Follow the principle of "CANI"—Constant
And Never-ending Improvement.
—Anthony Robbins

Effective project managers are keen to increase their team members' performance. They continually develop techniques and processes to make incremental improvements to meet project objectives. Project managers must involve project team members and associated stakeholders in developing techniques to improve performance. They must address the concerns and comments stakeholders raise in implementing such techniques and processes.

However, if project managers proceed unilaterally without gaining stakeholder acceptance and commitment about these techniques, stakeholders resent the process, possibly leading to negative politics. If project managers develop techniques and processes to make incremental improvements through genuine stakeholder participation, acceptance, and commitment, their efforts will be viewed positively, and stakeholders will more likely cooperate.

7.2.4. Forming coalitions—negative or positive impact

Call it a clan, collated network, call it a tribe,
and call it a family: whatever you call it,
whoever you are, you need one.

—Jane Howard

Forming coalitions— assembling groups of people—is a common political strategy in organizations to meet common objectives. In a project environment, some people form coalitions with other team members or with people from different departments to work together toward a common goal.

Sometimes, weaker members of one department form a coalition to have more power than that of stronger participants in influencing decisions. By increasing their strength in numbers, they create a perception that their viewpoint or approach is much better and should get more organizational support (Mannix 1993, 1–22; Cobb 1991, 1057–1079; Stevenson, Pearce & Porter 1985, 256–268).[6] Their actions can have a positive or a negative impact on the project.

We sometimes see coalitions, or groups of employees, as cliques of employees or a "good old boys' network." Politically, they help project managers included in this network get into senior management positions, but they are not so good for those excluded from the group. Sometimes, women have difficulty getting into senior management because they are often excluded from powerful networks (Burke & McKeen 1992, 245–283; Ragins & Sundstrom 1989, 51–88).[7]

Forming coalitions and acting as a group have a negative impact in the organization if the coalitions focus on satisfying their self-interests or the interests of a few group members rather than looking after the organization's objectives. The behavior is negative if the coalition or a group engages in activities to lower productivity of the team, department, or organization.

The impact is positive if the coalition works toward improving a process or system in the organization, and the coalition was formed to meet a particular set of organizational objectives. For example, if an organization has a few quality problems in their manufacturing department, a group might get together to resolve those quality problems.

Sometimes, organizations encourage departments to form groups to work as a team to meet a common objective to benefit the organization. Often, organizations provide adequate resources and support to such coalitions. Examples of coalitions with positive impacts include quality circles, mentorship programs, teams to solve technical and manufacturing problems, and teams to improve safety of equipment and personnel. Senior management should learn if such groups exist in the organization and their chief activities. Then, they can tactically influence their activities and support them by providing resources to meet important organization goals.

7.2.5. Churning (organizational amnesia)— negative impact

> *When a thing is done, it is done. Don't look*
> *back. Look forward to your next objective.*
> —George C. Marshall

Churning is agitating cream or milk to make butter by repetitive stirring. In this context, *churning* refers to a behavior of bringing back the same point repeatedly, even if the decision has been made against it. For example, a project manager discussed various options (A, B, and C) about a project issue in a meeting with all stakeholders. All options were evaluated, and the consensus was to reject option C outright and choose option B over option A for implementation because of its benefits to the organization. The proponents of options A and C had full opportunity to present their cases. After discussions, they agreed to go ahead with option B.

However, after the meeting, option A's proponents continue to push for their favorite during discussions with other stakeholders and lobby to justify that option B was not the best option, and it should not have been chosen. They play politics to influence other stakeholders to push for option A rather than option B, even when they agreed to support option B in the meeting.

In later meetings, they keep churning, that is, they argue option A is better than B and the previous decision of selecting option B was not a good decision. This political behavior is negative because it wastes time and frustrates other project team members and stakeholders. They fail to understand why some people do not cooperate and support the decisions made with their input and the input of most stakeholders. It can be described as "loss of memory" or "organizational amnesia," and the behavior is negative because it typically leads to negative politics.

People behave differently in a team environment from how they behave at an interpersonal level. Their overall performance in completing project assignments is evaluated by their performance at an individual level and at a team level. They might use political behaviors to gather information, control information, and filter information. They might form a coalition with other team members to increase their power as a group to push their favorite ideas or block decisions they do not like. Project managers and management should know these political behaviors to encourage positive behaviors such as commitments to make incremental improvements and take corrective actions to minimize the negative behaviors that might affect team performance and the overall project.

7.3. Political Behaviors Related to the Organization

> *Political behavior is a way of dealing with organizational conflict and differences.*
>
> —Anonymous

Management and project personnel might behave differently at an organizational level compared with how they behave at the interpersonal and project team levels. Management must select a portfolio of programs and projects to meet organizational strategies. Senior management is responsible for understanding different people's political behaviors and their impact on the organization's performance.

Various political behaviors related to the organization and their impacts are shown in table 7.3.

Behaviors Related to the Organization	Impact
Developing a Sense of Obligation	Positive
Managing Impressions	Negative/Positive
Forming Conclusions without Thorough Objective Analysis	Negative
Smoke Screening	Negative
Avoiding Making Decisions	Negative
Nominating for Rewards	Negative/Positive

Table 7.3. Political Behaviors and Impact Related to the Organization

7.3.1. Developing a sense of obligation—positive impact

> *In truth, people can generally make time for what they choose to do; it is not really the time but the will that is lacking.*
> —Sir John Lubbock

Developing a sense of obligation is an important skill in project management because it shows positive behavior. However, some project managers increase their power using the norms of reciprocity. As a political tactic, they help or do a favor for someone and create a sense of obligation to get something (even bigger) (Cohen & Bradford 1989, 5–17).[8] It requires developing good rapport and working relationships with project personnel so they are self-motivated and want to do their project assignments, thereby creating a sense of commitment and ownership among project personnel.

This skill is challenging, but once project managers master it, the dividends are high. The following factors should be considered to develop a sense of obligation and gain commitment from project personnel to meet project goals:

1. **Be involved genuinely** — Project managers should involve their team members and other stakeholders in defining and planning tasks. The stakeholders should be involved in developing processes and techniques outlining how the tasks are done. Involvement must be genuine, and project managers must listen to their stakeholders for comments and questions. If involvement is just lip

service, stakeholders do not feel motivated to complete the task effectively.

2. **Gain acceptance** — After involving stakeholders genuinely, project managers should strive to gain their acceptance for major project management activities, such as project planning and execution. Project managers must communicate clearly with stakeholders and seek their feedback. They should encourage stakeholders to raise questions and concerns and address those appropriately and promptly.

   Gaining acceptance is a prerequisite to creating a sense of obligation. Project managers should not enforce their ideas and decisions on project team members and stakeholders. Instead, they should try to create an environment in which the stakeholders agree on their own.

   To gain acceptance to help create a sense of obligation, you must ensure stakeholders carry through on their agreements. Three conditions must be met to increase this likelihood:

   1. The agreements must be free of any fear, threat, or coercion.

   2. The agreements must be fair and reasonable.

   3. There must be good, mutual understanding between the two parties.

Project managers must recognize that if any of the three conditions are not met, stakeholders are unlikely to follow through on the agreements. For example, any agreements made under fear or threat will be enforced without giving any opportunity to ask questions and raise concerns. Similarly, if any agreements are perceived unfair and unreasonable, stakeholders will think they are exploited and believe project managers are taking undue advantage of their power status.

The third condition is important in that if it is met, the other two conditions are easier to meet. This condition is more associated with establishing trust that leads to a meeting of the minds. It implies that when there is good understanding between two parties, they use open and honest communication and reach fair and reasonable agreements.

3. **Empowerment** — Explain assignments in *what*, and not necessarily *how*. Empowerment is critical to motivate project personnel and create a sense of obligation and commitment long term. *Empowerment* refers to giving autonomy to decide for planning, executing, and improving project activities. It involves explaining *what* must be done, and not *how* it should be done.

   The chief theme behind empowerment is that project managers must show confidence in their team members' ability to do their assignments effectively without micromanaging them. Project managers should support their team members in every possible way by giving them proper advice and feedback when and where the team members might need. When team members feel empowered, they accept challenges and go an extra mile to produce extraordinary results, not only improving overall project performance, but also motivating team members and increasing their commitment to meet project goals and objectives.

4. **Sense of pride** — This subtle, but powerful, factor creates a sense of obligation and increases commitment among project stakeholders long term. It involves believing in their ability, integrity, and high self-motivation. When people take pride in their jobs, they accept full ownership and put in extra effort to do a high-quality job.

   The sense of pride also leads to the following:

   a. Better understanding
   b. Innovation
   c. Increased cooperation
   d. High levels of ownership and commitment

These four factors are crucial to develop a sense of obligation among project personnel. These factors are part of important project management and leadership skills. Such skills are difficult to develop, but project managers must recognize that dividends of practicing these skills are high. Incorporating these factors in project management practice creates an environment of better working relationships and understanding for one another. The communication and negotiation processes are more open and honest, increasing self-motivation and commitment of project personnel to deliver successful projects. In addition, it increases mutual trust, understanding, and teamwork.

7.3.2. Managing impressions—negative or positive impact

> *Your expression is the most important thing you wear.*
> —Sid Ascher

Managing impressions means creating impressions that influence perceptions to meet objectives. It means acting to enhance our public image (Giacalone & Rosenfeld 1991; Tedeschi 1981).[9] Project managers and stakeholders might act specifically to create required impressions. As an impression management tactic, some executives manage their appearance and behavior to create a desired image (Gardner 1992, 33–46; Liden & Mitchell 1988, 572–587; MacGillivray, Ashcroft & Stebbins 1986, 127–135).[10]

Managing impressions might also involve changing the organization to convey new messages to its
customers and clients, such as going from national to
global, from retail to wholesale, from manufacturing
to only sales support. This change might require
major strategic changes or changing the company's
logo and website, physical location, appearance, and
so on.

Sometimes, managers and project stakeholders
use organizational resources to create impressions
just to meet their objectives and self-interests. Such
actions and behaviors are negative and lead to negative politics. For example, Paul Stern, former CEO of
Northern Telecom used image-building strategies to
"be distinct from the crowd" by wearing European-
tailored suits and using original Renaissance art in
his office to give an impression of an international
executive with a strong interest in high-quality and
expensive choices (Stern & Shachtman 1990).[11]

However, if these impressions are managed
for the organization's benefits, such actions and
behaviors are positive. CBC (Canadian Broadcasting
Corporation) management recognized that perception is a reality in business, and they saved their
budget cuts by improving public image and offsetting
the wrong impression.

7.3.3. Forming conclusions without thorough objective analysis—negative impact

*Before you begin, get good counsel; then
having decided, act promptly.*
—Sallust

Some senior managers play the politics of pushing their agendas to satisfy their self-interests. They give the impression of involving others and seeking their input to form conclusions and make final decisions. They create a perception that their decision-making process is logical and unbiased. However, their underlying objective is to push their ideas and skew the process to draw conclusions without objective analysis. This political behavior is negative because its chief purpose is to satisfy self-interests rather than the organization's interests, which creates significant frustration and unhappiness among stakeholders, leading to a lack of teamwork and cooperation.

7.3.4. Smoke screening—negative impact

A lie get gets halfway around the world before the truth has a chance to get its pants on.
—Winston Churchill

Smoke screening can be described as a technique of creating a screen to cover the truth—creating a dense cloud of smoke emitted to screen an attack or to cover a retreat. Some managers use this technique to prove that their positions and ideas are much better than others on the table are. They make presentations to prove their point by using statistics and anecdotal statements to their advantage. They influence other stakeholders to accept their ideas and cooperate in implementing those ideas. This political behavior is negative because some managers use it to push their favorite ideas and techniques, which might lead to wrong decisions for the organization.

7.3.5. Avoidance in making decisions—negative impact

Take time to deliberate, but when the time for action has arrived, stop thinking, and go on.
—Napoleon Bonaparte

Making good and timely decisions is critical for project success. Project managers want to have full information and input from project team members and stakeholders before making a final decision. They postpone deciding because of lack of confidence, lack of information, and insecurity. However, in some situations, project managers need more time to gather desired information and input, but they have to decide with available information.

In such circumstances, project managers behave in many ways, depending on their self-confidence and abilities to decide. Some managers follow "malicious compliance"; that is, they decide in their favor and give the impression to stakeholders that those decisions are necessary to comply with regulations and protocols. They decide to "do nothing" to protect themselves and argue that it is better to make no decision at the time to be consistent with organizational policies and procedures. They might do the following:

- Avoid deciding
- Delegate decision-making to someone
- Delay in deciding
- Use malicious compliance

Most of these techniques are negative political behaviors because they lead to frustration and tension among project team members and stakeholders. Project managers use these techniques in these circumstances:

- Lack of abilities to analyze and manage risks
- Lack of self-confidence and feelings of insecurity
- Lack of relevant and authentic information
- Lack of time to evaluate available information and input

Organizational success depends on decision quality and timing. Good decisions are made promptly, and they help meet project objectives and organizational goals. During the planning stage, project managers must identify risks and develop strategies to manage those risks. The decision-making processes must consider the unexpected problems and constraints related to human factors, economic factors, and project schedules.

Instead of using negative political behaviors, project managers must recognize their weaknesses and develop appropriate skills and confidence in making good decisions promptly. They must learn to gather and evaluate relevant and authentic information efficiently to make good decisions, which helps them gain leadership skills to create a climate of better collaboration and teamwork.

7.3.6. Nominating for rewards—negative or positive impact

Rewards and motivation are an oil change for project engines. Do it regularly and often.
—Woody Williams

Nominating for rewards is so common that it might be perceived as political. Most people like rewards because they satisfy the needs they value most. Some rewards are won through nominations, which increase visibility and profile, so people might use politics to be nominated for such rewards. They use their close contacts and friends to nominate them. Some people like to nominate their friends to get favors in return from them. This nomination process can be positive if it is done objectively, and the people nominated deserve the awards. It is negative if it is subjective, and people play political games to help only their close friends for their self-interests.

Management and project personnel behaviors can differ at an organizational level compared with those at the interpersonal and project team levels. Organization-related behaviors and their impact include developing a sense of obligation (positive), managing impressions (negative or positive), forming conclusions without thorough objective analysis (negative or positive), smoke screening (negative), avoidance in making decisions (negative), and nominating for rewards (negative or positive).

Organizational leaders should know various political behaviors displayed by senior managers and project managers and their team members because such behaviors affect the organization. Senior management should encourage positive behaviors and take corrective actions to minimize impacts of negative behaviors on overall organizational performance. Similarly, project managers should understand behaviors and the motives behind them. They should analyze and evaluate the impact of all political behaviors on project outcomes and take necessary actions to encourage the positive behaviors and discourage behaviors that lead to negative politics.

Chapter 7 Summary

Political behaviors can be divided into three categories: 1) political behaviors at the interpersonal level, 2) political behaviors related to the team environment, and 3) political behaviors related to the organization. Political behaviors at the interpersonal level include finger-pointing (attacking or blaming others), name-dropping, introducing change, networking (establishing contacts and relationships), and sharing of information. Networking is a positive behavior, especially if it is used to help the organization, whereas finger-pointing or blaming others is a negative behavior and does not lead to any productive outcomes. The other three behaviors at the interpersonal level can be positive or negative depending on the motives behind them.

People behave differently at the interpersonal level to manage uncertainties, difficult situations, and politics in project environments. People should be sensitive to others' feelings, and they should treat others the way they want to be treated. Project managers should know their project personnel's behaviors at the interpersonal level, understand their motives, and encourage positive behaviors to minimize the impact of negative behaviors.

Political behaviors related to the team environment include controlling information, filtering information, incremental improvements, forming coalitions, and churning. Actions aiming toward incremental improvements are considered positive because they are intended to increase the organization's productivity. The remaining four behaviors related to the team environment can be both positive and negative depending on team members' strengths and attitudes.

People behave differently in a team environment from how they behave at an interpersonal level. Their overall performance in completing project assignments is evaluated by their performance at an individual level and at a team level. They might use political behaviors to gather information, control information, and filter information. They might form a coalition with other team members to increase their power as a group to push their favorite ideas or block decisions they do not like. Project managers and management should know these political behaviors to encourage positive behaviors such as commitments to make incremental improvements and take corrective actions to minimize the negative behaviors that might affect team performance and the overall project.

Political behaviors related to the organization include developing a sense of obligation, managing impressions, forming conclusions without thorough objective analysis, smoke screening, and nominating for rewards. Developing a sense of obligation among stakeholders increases their levels of commitment and motivation, leading to positive results, whereas smoke screening is considered negative because it is done mainly to create perceptions about the superiority of your positions and ideas.

Organizational leaders should know various political behaviors displayed by senior managers and project managers and their team members because such behaviors affect the organization. Senior management should encourage positive behaviors and take corrective actions to minimize impacts of negative behaviors on overall organizational performance. Similarly, project managers should understand behaviors and the motives behind them. They should analyze and evaluate the impact of all political behaviors on project outcomes and take necessary actions to encourage the positive behaviors and discourage behaviors that lead to negative politics.

The behaviors in these three categories can be positive or negative. Project managers should understand these behaviors and the motives behind them. They should analyze and evaluate the impact of all political behaviors on project outcomes and take necessary actions to encourage the positive behaviors and discourage the behaviors that might lead to negative politics.

The Art of Positive Politics

Managing Political Behaviors

We cannot safely leave politics to politicians,
or the political economy to college professors.

—Henry George

Politics are inevitable in project environments because project participants come from different departments, organizational cultures, diversities, and interests. Project stakeholders behave in different ways to do things. Sometimes, they indulge in politics to meet their objectives or their project objectives, which might lead to negative impacts on other programs and projects. Negative politics create enormous anxiety among people, often motivating them to take actions to meet personal agendas, sometimes at the cost of organizational goals.

Senior management can take primary initiatives that help minimize the negative impact on organizational and project politics. These initiatives must deal with issues and problems, which, if not properly addressed promptly, create negative politics. Therefore, senior management must develop appropriate initiatives and implement policies and processes to ensure everyone in the organization follows the initiatives. Such major initiatives can be divided in two categories to deal with major issues and concerns at the interpersonal and team levels and at the organizational level (Verma 2014; McShane 2014; McShane 1995, 355–375).[1]

8.1. Initiatives at the Interpersonal and Team Levels

> *Coming together is a beginning; keeping together is progress; working together is success.*
>
> —Henry Ford

Projects are done by project teams consisting of people with different personalities, attitudes, and skills. Wherever there are people, there are politics. Project managers and their team members might behave in certain ways to minimize their stress, anxiety, and risks in dealing with project politics and organizational politics. Good management practices should be developed and followed to manage negative political behaviors at the interpersonal and team levels. Initiatives and guidelines for managing political behaviors related to interpersonal and team levels are shown in table 8.1.

Guidelines	Must Develop and Follow
Encourage sharing information.	Good management practice
Establish good organizational values and team norms to discourage negative politics.	Good management practice
Establish ground rules and enforce them to discourage negative politics, including the following: • Managing meetings (agenda, responsibilities, action items, and so on) • Conflict management and escalation • Brainstorming and facilitation • Problem solving and decision making	Good management practice
Consider cultural aspects.	Good management practice

Table 8.1. Guidelines for Managing Political Behaviors at the Interpersonal and Team Levels

8.1.1. Encourage sharing information

> *Innovation comes only from readily and seamlessly sharing information rather than hoarding it.*
> —Tom Peters

In most organizations, information is power. People having more information themselves or having easy access to information about the organizational policies and procedures, project selection, and prioritization processes and systems are powerful. Some people hoard or hide information to maintain their power. Technical specialists who feel insecure about their position and future use this technique to demonstrate their power. The information is an informal power, and the more you share, the more you get.

It is important to disseminate information effectively and promptly and discourage people from withholding information to gain power. Good senior management creates an environment where sharing information is not only encouraged, but also rewarded. This environment empowers and motivates people working on programs and projects to share their information more widely by putting on company intra sites, brown-bag lunches, presenting internal seminars and workshops on lessons learned, and so on.

8.1.2. Establish good organizational values and team norms to discourage negative politics.

> *It takes each of us to make a*
> *difference for all of us.*
> —Jackie Mutcheson

People engage in politics when there are no good organizational values and team norms to direct them. If senior management does not practice good values that encourage sharing of information, open communication, and objective feedback, employees play politics to get what they want. In addition, if team norms are not based on good values to focus on synergy, providing mutual support and helping one another by sharing information, some team members do what advances them rather than the team, leading to negative politics.

Lack of understanding in following good organizational values and team behavior is a common reason people indulge in negative politics. Senior management should develop good organizational values and team norms to increase ethical standards and empower project teams to produce extraordinary results.

- **Establish good organizational values —** To achieve this, senior management should take these steps:

  - Lead by example (high integrity and ethical standards).

  - Create an organizational mission and strategies that provide an uplifting and noble future.

  - Be transparent and objective in setting priorities and distributing rewards.

  - Respect and listen to people to empower them and nurture creativity.

  - Provide challenging opportunities and proper organizational support at all levels.

  - Emphasize growing people at all levels (professional, moral, and personal).

  - Be socially responsible and urge the staff to do so.

- **Establish good team norms —** To achieve this, senior management should use these guidelines and encourage project managers to do the same at the project team level:

- Separate people from problems and focus on interests, not positions.

- Evaluate ideas without judging people. Once an idea is tabled, it should not be criticized, because it now belongs to everyone.

- Encourage sharing information to achieve cross training and high efficiency.

- Create high-performance teams by providing them with challenges, supporting them, and nurturing creativity and innovation.

- Strengthen the weakest link in the team because teams work interdependently, and performance of one team member depends on the performance of another. It can be done by taking the following steps:

  - Emphasize unity of purpose and synergy.

  - Base rewards and recognition on objective criteria. Rewards in general should be given to the team, and the team should decide any individual rewards.

- Project managers should follow this guideline: "When it comes to taking credit, give it to your team; when it comes to taking blame, take it upon yourself." —Anonymous

278

8.1.3. Establish ground rules and enforce them to discourage negative politics.

The very essence of all power to influence lies in getting the other person to participate.
—Harry A. Overstreet

Senior management should discourage negative politics that include pushing personal agendas, gaining control, and working in silos. This book's purpose is to instruct you in how to understand politics and manage them effectively. Program and project managers should be proactive and try to discourage negative politics, starting at their team level, and then trying to influence upper management to do the same. They should create an environment to encourage teamwork and synergy, helping one another win, and cross training to improve all team members' project management competencies. They should reward people for good behavior and discourage unsupportive and disruptive behavior of team members.

Some team members do not behave professionally when dealing with other team members. The following are major areas that might lead to negative politics and behavior that might lead to lack of mutual trust, respect, support, and cooperation. Project managers should establish ground rules, processes, and systems to curtail unproductive behavior associated with these areas to manage their projects effectively and enhance the performance of all team members.

- **Managing meetings** (agenda, responsibilities, action items, and so on) — Project personnel spend much time in different meetings throughout a project's life span. Project managers often get frustrated with a few team members' disruptive behavior, leading to less cooperation and an unpleasant team environment. Project managers should establish a few ground rules and clarify their expectations associated with managing various meetings with different stakeholders. Once a decision is made and agreed upon, stakeholders should not revisit the decision because it wastes time. Important issues include the following:

- Conducting meetings and defining and following up on action plans

- Preparing for meetings

- Sharing ideas

- Preventing monopolization of discussions

- Respecting all team members

- **Conflict management and escalation** — Conflict management and escalation are areas that lead to problems in project environments because projects are done by people with different backgrounds, expectations, interests, and norms. Conflicts are inevitable, and project managers must manage them promptly.

  Project managers should develop a general conflict management framework and process to resolve conflicts quickly as they happen.

They should recognize that the longer the conflicts are left unresolved, the more problems will arise. They should develop a process to escalate problems promptly to appropriate levels. Project managers must analyze the conflict situation, consider these factors, and then use the appropriate conflict management strategies:

- Is the conflict real or a misunderstanding?
- What is the nature of the parties involved?
- What is the timing?
- What are the chief interests and concerns of both parties?
- What is the emphasis from both parties on goals and relationships?

Project managers should use one of the following five conflict management strategies after considering the factors and any other relevant information available to them:

1. Avoidance (withdrawal)
2. Forcing (dominating)
3. Accommodating
4. Compromising
5. Problem-solving

- **Brainstorming and facilitating** — Project managers must facilitate various technical project requirements, such as project planning and estimating sessions, change request systems, and risk identification and analysis. Team members should also be encouraged to devise new ideas that must be analyzed and evaluated objectively. Some team members are too aggressive and criticize new ideas and plans without proper analysis and evaluations. When team members criticize people rather than their ideas, negative impacts can occur. The following are key points to remember while conducting a facilitating session:

- All ideas presented should be recorded using words identical to the words of the idea originator. It gives them the feeling their ideas have been listened to, acknowledged, and recorded.

- Once any team member presents an idea, he or she should not feel too attached to or emotional about the idea.

The team must analyze and evaluate ideas to make a final decision whether to accept or reject the idea, based on objective and logical analysis. Both evaluation and the final decision affect the feelings of the person who originally presented the idea, depending on the person's emotional attachment to the idea.

Team members should separate the person from the idea. Then, the idea belongs to the group, and not to the originator only. Therefore, any evaluation and criticism of the idea should be directed at the idea, and not at the originator. The group will either accept or reject the idea.

In the end, the idea originator should not be upset with either outcome, even if the idea is rejected, if the team members do not attack or criticize the idea's originator. It is even more supportive and good management practice if senior management recognizes the person for the idea the team accepted. This practice encourages more creativity and innovation from team members.

Project personnel are likely to get upset if aggressive team members attack, block, or reject their ideas right away with no logical explanation. In such environments, people play politics to push their ideas. They might try to form coalitions to gain wider acceptance of their ideas or viewpoints.

Project managers can curtail such unproductive environments by establishing good ground rules for brainstorming and facilitation that encourage more creativity and innovation. Ground rules should discourage people from attacking others. The purpose of these ground rules is to continually improve overall personal and team performance. Project managers should separate people from ideas. They should evaluate ideas objectively for practicality and ease of implementation.

• **Problem-solving and decision-making** — In problem-solving and decision-making, some project personnel often think politics are used for pushing personal agendas. Sometimes, people with more power and connections use politics to get decisions made in their favor. Similarly, a few selective people decide the final solutions for some problems or issues, and many project personnel are frustrated. The final decisions might benefit a few people personally rather than help the organization. Many project personnel feel frustrated because the decisions were not made objectively, and they were not given any opportunity to present their viewpoints or debate the issues associated with final decisions.

To curtail such activities, project managers should establish ground rules explaining and clarifying to all team members how decisions are made. Management should ensure that procedures are created after thorough objective analysis and input from relevant stakeholders. The objective criteria should be defined and followed to minimize the anxiety of people and negative impact of politics.

8.1.4. Consider cultural aspects

People have one thing in common;
they are all different.
—Robert Zend

Most projects are becoming global, and culture is an important factor in multinational projects. The literature on culture and cultural dynamics suggests that cultural diversity and international project management are inseparable (Verma 2014; McShane 2014; Verma 1997, 89–110; McShane 1995, 355–375).[2] A thorough understanding of the impact of culture on teamwork, leadership, and communication is the key to project success in a global economy. Cultural values drive behaviors, attitudes, and actions. Any differences in behaviors and actions can be misinterpreted as intentional actions to play politics, whereas they might be merely driven by cultural differences and orientations (Verma 2014; McShane 2014; McShane 1995, 355–375).[3]

Culture can be defined as a commonly shared set of values, beliefs, attitudes, and knowledge that can be transferred from one generation to the next through family, school, social environment, and other agencies (Hofstede 1993, 5; Martin 1981).[4] It represents a particular set way in which a group of people associated by common religion, geographical region, ethnicity, or strong principles lead their lives (Terpstra 1978, 2, 176; Terpstra 1972, 83).[5] Culture is complex and difficult to define because of its several elements and dimensions.

Project managers should respect cultural differences and capitalize on cultural levels. According to Hofstede (1993, 5), culture is a mental software . . . the collective programming of the mind that distinguishes members of one group from another.[6] Culture can also be defined as an acquired knowledge to interpret the experiences of the group that led to its behavior (Luthans & Hodgetts 1991, 35).[7]

Cultural diversity is an important part of the project environment to survive in a global economy. Global business organizations must urge their project managers to develop intercultural communication and negotiations skills to manage international projects. Project managers must identify the major elements of culture, critical dimensions of cultural differences, and cultural variables as follows (Verma, 2014; Verma, 1997, 89–110; Martin, 1981):[8]

- **Major elements of culture** — Martin identified the following seven major elements of culture:

  1. Material culture

  2. Language

  3. Aesthetics

  4. Education

  5. Religion, beliefs, and attitudes

  6. Social organization

  7. Political life

  Project managers must understand the elements of culture to manage international projects successfully.

- **Critical dimensions of cultural differences** — Hofstede (1993, p. 14), Turner & Trompenaars (1993) and Owens & McLaurin (1993, 229–236) identified major critical dimensions of cultural differences:[9]

  - Power distance index (low and high)

  - Individualism versus collectivism

  - Uncertainty avoidance

  - Gender role identification (masculine or feminine)

- Time (time horizon, the value of time, and focus)

- Attitude toward life (quality versus quantity of life)

Project managers must appreciate the critical dimensions of cultural differences to achieve synergy in a multicultural work environment.

- **Major cultural variables and associated orientations** — Rick Punzo described five major cultural variables with associated orientations that affect the behavior of people in different cultures (Punzo 1996, 863–868; Training Management Corporation 1996):[10]

  1. Environment: Environment influences the behavior of people with respect to the orientations of harmony, control, and constraint (Hall 1990; Kluckhohn 1961).[11]

  2. Action: Cultures can ascertain whether people are aggressive or passive and by how much. People's orientation to action can be task-oriented or relationship-oriented (Hall 1990; Kluckhohn 1961).[12]

  3. Communication: Different orientations to communication can lead to breakdown in communications and problems in project management, sales, negotiations, and teamwork. Cultural orientations to communication can be formal or informal, direct or indirect, low context or high context, expressive or objective (Hall 1990; Kluckhohn 1961).[13]

4. Space: Different cultures have different preferences about how they use their space while communicating with others. The cultural orientation to space can be either private or public (Punzo 1996; Rhinesmith 1993; Hall 1990).[14]

5. Thinking: *Thinking* refers to how something is interpreted and analyzed. Cultural orientation to thinking can be inductive, deductive, linear, and systemic (Punzo 1996; Rhinesmith 1993; Stewart & Bennett 1991; Hall 1990).[15]

Project managers should develop awareness of these cultural variables and their associated orientations to capitalize on cultural differences and achieve synergy and cooperation among team members of different cultures.

Successful project management is the key to organizational success, and managing a project team is critical to deliver successful projects. People behave in certain ways at the interpersonal level and at the team level to protect themselves and gain support from others.

To manage political behaviors at the interpersonal and team levels, encourage sharing information, establish good organizational values and team norms to discourage negative politics, establish ground rules and enforce them to discourage negative politics, and consider cultural aspects. All these guidelines must follow good management practice.

8.2. Initiatives at the Organizational Level

*Organizational structures alone do not
determine project success; what matters is the
commitment to make them work.*

—Anonymous

Today, many organizations use Management
by Projects (MBP) as an approach to manage their
overall organization. Politics are a necessary part
of life in project environments. People at different
levels in the organization use different political
behaviors to increase their power and influence to do
things. Several political behaviors affect the organization positively or negatively. These behaviors are
associated with the project selection process, organizational structure to avoid centralization of power
among few, a fair and objective reward system, and
good hiring practices.

Sound policies should be established and followed to manage negative political behaviors at
the organizational level. Senior management must
develop suitable initiatives to resolve issues at the
organizational level and implement proper policies
and procedures to ensure everyone in the organization follows them. Initiatives and guidelines for
managing political behaviors related to issues at the
overall organization level are shown in table 8.2.

Guidelines	Must Develop and Follow
Establish appropriate policies and good management practices related to the following: • Project selection • Project prioritization • Allocation of scarce resources	Management policies and good management practices
Avoid centralizing power among a few to avoid possible power corruption.	Good management practices
Hire senior managers with high moral values.	Good management practices
Give rewards and recognition based on objective criteria.	Management policies and good management practices

Table 8.2. Guidelines for Managing Political Behaviors at the Organization Level

8.2.1. Establish appropriate policies and good management practices

> *Getting your house in order and reducing the confusion gives you more control over your life. Personal organization somehow relates or frees you to operate more effectively.*
>
> —Larry King

Lack of an objective process to select projects is a common issue senior management should deal with to avoid creating negative politics. Without logical directions and policies, much confusion at all organizational levels leads to lack of teamwork and cooperation. To deal with this properly, senior management must establish proper rules based on objective criteria, and then enforce them. It works better if appropriate parties are involved in this process, and rules are communicated to everyone affected by them to gain acceptance and commitment from maximum stakeholders. These rules aim to primarily resolve problems related to scarce resources that typically result in politics. Project managers play politics to get the best resources to meet their project objectives.

These rules should address several issues:

- **Project selection** — In many industries and organizations, employees from research and development departments and other functional areas create ideas and projects. Not all those projects can be done because of limited financial, labor, and other resources. In software and IT industries where many more ideas and projects are initiated than the organization can manage with its capacity, these projects should not even start.

  However, this often happens because the sponsors of such projects start with low estimates of required resources, budget, and time for their favorite projects, and they are politically strong enough to push those ahead, even at the cost of other important projects. It leads to a risk that projects important to meet organizational strategies might not get the appropriate resources, especially if their sponsors or champions cannot manage organizational politics to get their projects in the overall portfolio of organizational programs and projects.

Often in IT industries, many ideas are generated for many projects that should go through an objective review process. Then, only one-tenth of those ideas are selected, and relevant projects are defined to meet those ideas, much like putting ideas through a funnel of objective review to select the good ones. It provides a process of comprehensive objective analysis and evaluation before projects are selected. The difference between projects and tasks should be understood in terms of the following:

- Size of projects in financial, manpower, and other resources required
- Impact on organizational strategies
- Active requirements from cross-functional departments

This is important because some people in organizations use politics to push their ideas that can be organized as a task or a small work package rather than a project, leading to inappropriate allocation of resources.

- **Project prioritization** — Most project managers want to work on high-priority and high-profile projects because it is easier for them to get required resources. Otherwise, many project managers indulge in politics, trying to influence the decision-makers who assign priorities to various projects in the organization.

Sometimes, a particular project should be given a lower priority and should not even be started because it requires more resources and a special skill mix than the organization can allocate, based on its capacity. However, the politically strong sponsors of such projects start with low estimates of required resources and start the project. Once the projects proceed, they experience significant time delays and cost overruns, even in the early stages of the project life cycle. The sponsors play politics and argue to continue with these projects because of substantial investment made already.

If such projects are not crucial to meeting organizational strategies, senior management should minimize their losses and terminate such projects to avoid wasting more resources that could be assigned to true high-priority projects.

The political maneuvering that leads to anxiety happens under the following situations:

- There are no clear rules based on objective criteria to establish priorities.

- If the rules are there, they are not communicated clearly to various functional managers, program managers, and other related stakeholders.

- Even if the rules exist, they keep changing without explanation, and the project managers are often the last to learn these changes. When they do, it is only through the grapevine and their informal network.

The negative impact of politics can be minimized if senior management establishes a system to decide on priorities based on objective criteria for various projects and then ensures the resources (especially the scarce and specialists) are allocated accordingly.

*   **Allocation of scarce resources** — Lack of resources, especially the ones with special skills, is a major factor leading to politicking, favoritism, and game playing. After a few years of practical project management experience, most project managers recognize that projects are done by people, not by computers or software packages.

    Most projects are organized in a matrix whereby project managers estimate human resources and then negotiate with functional managers to get the best resources for their projects. They want the resources with the most appropriate skill level to match various work packages and tasks. They try to influence upper management, functional managers, and others in the organization who make final decisions about the allocation of resources.

The chief reason for negative politics in most organizations is the lack of clearly communicated processes and systems to select a few projects in the overall portfolio of programs and projects that align with the most important organizational strategies and goals. The worst situation is that, sometimes, the decisions to select projects, assign priorities, and allocate resources are ambiguous and subjective and keep changing without proper reasons and analysis. Significant organizational resources might be inappropriately used if such processes and systems are not based on objective criteria and then not clearly communicated to the managers responsible to execute the programs and projects on time and budget and within organizational constraints.

Senior management should emphasize the importance of organizational strategies and goals and should create a work environment where people focus on spending their energy and efforts to meet organizational goals and strategies rather than worrying about their goals.

8.2.2. Avoid centralizing power among a few to avoid possible power corruption

> *Power tends to corrupt, and absolute power corrupts absolutely.*
> —Lord Acton

Power often leads to politics, and the way in which power is exercised determines whether the politics have negative or positive impacts. Departments and individuals have more power as their centrality increases. *Centrality* refers to the degree and nature of interdependence between the power holder and others (Brass & Burkhardt 1993, 441–470; Hackman 1985, 61–77; Hickson et al. 1971, 219–221).[16] If power resides in the hands of only a few people, power corruption is possible, which often happens in dictatorships and autocratic and authoritarian environments. For this reason, in many organizations, hierarchical structures are flattened to distribute power among many departments and people.

When only a few people hold power, they might make a coalition to carry out activities to help their departments more than the overall organization. By distributing power appropriately, management can reduce the possibility of power corruption because various departments must be accountable to work together as a team to meet overall organizational goals and objectives.

8.2.3. Hire senior managers with high moral values.

You are free to choose, but you are not free from the consequences of your choice.
—Anonymous

High moral values are important at all organizational levels. However, it is more important that people at the senior management level have high moral values because they are expected to lead by example. A wrong person in senior management can create many serious problems, and therefore, it must be avoided. Senior managers should set examples with high moral values and develop a sense of responsibility among stakeholders. The following are common techniques to ensure senior managers hired have high moral values:

- Use attitudinal and situational questions and analysis besides interviews.

- Follow formal reference checks. (Be cautious of missing information because of privacy acts.)

- Gather information using informal networks, for example, golf courses, tennis clubs, sports associations, and cultural associations, because potential candidates for senior management positions might belong to the same clubs and associations.

8.2.4. Give rewards and recognition based on objective criteria

It is no good saying you can't afford to look
after your staff, you can't afford not to.
—Julian Richer

Rewards are important to motivate people. Rewards are most effective when they satisfy the needs people value the most. However, unfair distribution of rewards and recognition can be equally damaging to people's morale in any organization. Many project personnel commonly believe there is much favoritism in deciding to give rewards and recognition.

Many people get frustrated when some people are rewarded and recognized based on whom they know in powerful positions rather than based on their performance. Organizations should have clearly defined processes and systems that are communicated and followed to distribute rewards based on objective criteria rather than using subjective criteria.

Sometimes, even if systems are defined, there is still much favoritism, and many people consider it negative organizational politics. Often, even nominating someone for important rewards and recognitions (especially in professional associations and large projects) is considered prestigious because people are well profiled and promoted during the process to give rewards to these professionals or their teams. Reciprocating favors is common in this process.

Everyone likes rewards because they satisfy people's needs and motivate them to do better. Bob Nelson (1997) suggested 1,001 effective ways to energize employees in organizations.[17] However, the policies and process to distribute rewards and recognition must be fair and based on an objective criterion. An unfair reward system encourages people to get involved in negative politics.

Some senior management play favorites in giving rewards or even nominating their favorites for organizational rewards and recognition. Sometimes, it is done in a reciprocating system, that is, one senior manager nominates someone from a department for important rewards, and then the other manager reciprocates. Even the nomination process and the associated publicity increase the profile of the selected people and programs, even if they don't get the reward or recognition.

Problems happen when the system becomes subjective, leading to favoritism and negative politics. Again, such political behavior can be curtailed by establishing a reward and recognition system based on objective criteria of high performance to meet organizational goals and objectives. Policies should be fair and based on objective criteria, then clearly communicated to the managers responsible and involved in deciding rewards and recognition. Good leadership and efficient management are the keys to organizational success. The political behaviors displayed by senior management, project managers, and team members must be understood to evaluate the impact on individual and overall team performance. Senior management should use the guidelines outlined here to increase organizational performance at all levels.

Senior management should have a high emotional quotient (EQ) to be an effective leader and high moral values to lead by example. They should establish flatter organizational structures to avoid centralizing power among a few to avoid power corruption and establish an objective process to select the projects, prioritize the projects, and allocate resources. Reward systems should be based on objective criteria and performance to meet organizational strategies and goals.

Chapter 8 Summary

Politics are not inherently negative or positive, but depend on how you approach them. Politics are inevitable because teamwork requires the cooperation of people from diverse backgrounds, each with personal opinions and aspirations. Therefore, you must recognize how some behaviors, though often seemingly harmless, are politically motivated. Negative politics arise when employees blame others, withhold, filter, twist, or otherwise misrepresent information critical to project goals. They also occur when employees centralize power to pursue hidden agendas. Failure to make decisions promptly is another negative political behavior that delays completion of project goals. Negative politics lead to failures in efficiency, transparency, and employee morale.

Many political behaviors are often encountered in a project management environment. Understanding these behaviors is important, as it enables project managers and senior management to direct their project personnel's energy so it emphasizes a collaborative and team-oriented approach. Senior management should establish organizational values and team norms that discourage negative politics. Power should also not be centralized among a few managers and certain departments to avoid possible power corruption.

You should establish rules and policies that prevent team members from using their power and changing the project agenda to suit their goals. Establishing these policies also creates a framework through which project and organizational goals are clearly defined, and everyone's efforts are focused to meet overall organizational strategies. Positive politics center on positive reinforcement and clear understanding of project goals. Positive politics require employees to effectively problem-solve, accommodate one another, and work cohesively to achieve project goals on time.

To manage political behaviors at the interpersonal and team levels, senior management should encourage sharing information, establish good organizational values and team norms to discourage negative politics, establish ground rules and enforce them to discourage negative politics, and consider cultural aspects. All these guidelines must follow good management practice.

Senior management should try to analyze and manage political behaviors at the organizational level by developing and following appropriate management policies and good management practices to minimize the negative impact of politics. These behaviors include policies related to project selection, project prioritization, and resource allocation; avoiding power centralization among a few to reduce power corruption; hiring managers with high moral values; and giving rewards based on objective criteria. People get upset when appropriate policies are not established and good management practices are not followed because of the following:

• Irrationality

• More subjectivity

• Favoritism

• Lack of thorough analysis and evaluation

• Lack of involvement of appropriate stakeholders

• More emphasis on personal benefits than organizational benefits

• Lack of clear communication of policies and processes for giving rewards

Senior management should involve appropriate stakeholders in developing the policies and processes to gain their acceptance. Any changes should be communicated immediately, including an explanation why they were made. Rules and processes are not of great benefit if they are unfair, inconsistent, and not enforced.

Part III Summary

Project managers in any organizational environment often encounter three political positions. Some managers are politically naive and think politics are unpleasant and a waste of time, and they should be eliminated. Politically naive people can be compared to Flounders that stay passively on the ocean bottom with their eyes closed. Similarly, politically naive people ignore and avoid politics, and therefore, they often become victims of politics. These people trust others too readily, making them susceptible to harm from people who use negative politics.

The second type of managers can be compared with Sharks, as they try to manipulate people and processes to their advantage. They look at politics as an opportunity to advance themselves, even at the cost of others or organizational objectives. Like their namesake, they are predatory, in that they do not feel bad if others are hurt by their actions or face problems in managing their projects. With this self-centered behavior, Sharks alienate other workers, and they are unlikely to be rewarded with desirable opportunities.

The third type of manager is politically sensible. They can be compared with Dolphins because they are fast thinkers and agile. Politically sensible people have good intuition and people skills, and they are great in navigating organizational politics. They can analyze the political landscape quickly and identify power brokers and their use of power and influence. They can anticipate political turbulence to survive and thrive in a politically charged work environment.

Politically sensible managers also know the typical outlook of Sharks who engage in negative politics. They try to find good champions with high respect and positional power to protect themselves from Sharks. They enhance their informal power by building trusting networks, earning loyalty from their teams, and achieving organizational goals. Politically sensible people recognize the importance of people and depend on their human and interpersonal skills to deliver successful projects. They use technical skills to develop logical arguments and use effective communication skills to influence their stakeholders.

They are enthusiastic contributors and team players. They increase their informal power and help increase their team members' informal power. They build honest and long-term relationships and alliances. They continuously nurture and feed their networks to enhance the quality. They are trustworthy and know well how organizational politics work and how to interact with people with different personalities.

Upper management knows politically sensible people and their teams, and they are happy to reward them by recognizing them. This increases upper management's informal power and results in a positive feedback mechanism in which informal power, trust, and network continuously grow.

Part III also focuses on various political behaviors and their impacts (positive or negative) in a working environment and practical techniques to manage those behaviors. These political behaviors can be associated with three main levels or environments: interpersonal level, team environment, and the organization.

Politics are not inherently negative or positive, but depend on how they are approached. Politics are inevitable because teamwork requires the cooperation of people from diverse backgrounds, each with their opinions and aspirations. Therefore, it is crucial to recognize how some behaviors, though often seemingly harmless, can be politically motivated.

Negative politics arise when employees blame others, withhold, filter, twist, or otherwise misrepresent information critical to project goals. They might also occur when employees attempt to form coalitions to increase their power to pursue personal agendas. Failure to decide promptly is another negative political behavior that delays completion of project goals. Negative politics lead to failures in efficiency and transparency, and they lower employee morale.

Understanding and managing project stakeholders' political behaviors is important, as it enables project managers and senior management to direct their project personnel's energy toward a collaborative and team-oriented approach. Effective management of political behaviors involves taking initiatives at the interpersonal and team levels and at the organizational level.

Senior management should establish organizational values and team norms that discourage negative politics. In addition, power should not be centralized among a few managers and departments to avoid possible power corruption. Establish rules and policies that prevent team members from using their power and changing the project agenda to suit their goals. Establishing these policies also creates a framework through which project and organizational goals are defined, and everyone's efforts can be focused to meet overall organizational strategies.

Positive politics center on positive reinforcement and clear understanding of project goals. Positive politics require employees to effectively problem solve, accommodate one another, and work cohesively to achieve project goals promptly.

Part III Summary

Part IV. The Art of Managing Politics

Chapter 9: Three Truths of Life to Manage Stakeholders

9.1. Truth #1: People Make or Break Things

9.2. Truth #2: People Do Mostly What Is in Their Best Interest

9.3. Truth #3: People Support What They Create

Chapter 9 Summary

Chapter 10: Managing Politics at the Management Level

10.1. Organizational Issues

10.2. Leadership Issues

10.3. Project Management Issues

Chapter 10 Summary

Chapter 11: Managing Politics at the Project Level

11.1. Political Issues and Challenges of Managing Upward

11.2. Project Management and Team Leadership Issues

11.3. Stakeholder Management Issues

Chapter 11 Summary

Part IV Summary

The Art of Managing Politics

One of the penalties for refusing to participate
in politics is that you end up being governed
by your inferiors.

—Plato

Project management and politics are closely linked. Politics are a way to get things done in any organization. After understanding the dynamics of politics, the two types of politics (negative and positive), the political landscape, and various political behaviors, you must learn to manage politics to deliver successful results.

Politics exist throughout the project life cycle. During the conceptual phase, politics are important to finish the feasibility study and financial estimates to get funding for a multimillion-dollar project. Here, all internal and external stakeholders and their interests (personal and organizational) must be identified, and then, the feasibility study and conceptual plans must address What's in It for Them (WIIFT) to gain their acceptance and buy-in.

The Art of Positive Politics

During the detailed planning and execution phases, project managers must understand the dynamics of politics to facilitate and implement the planning and execution processes to get the work done from project team members from various disciplines across the organization. It involves communicating with stakeholders from different departments who do things in ways they believe best. Project managers must develop good communication skills and, especially, active listening, with special emphasis on political correctness. Project managers must sell project deliverables by emphasizing project benefits for each stakeholder and the organization.

Most projects have stakeholders with different viewpoints, expectations, and interests. In Part IV, three simple and logical universal truths of life are described. Although they seem simple, these truths are challenging, and they emphasize how important people are, what motivates them to do things, and how to gain their cooperation and support long term.

Wherever there are people, there are politics. Every organization has a good deal of politics at different levels. The level of politics is based on project complexity. The more complex the project is, the higher the intensity of politics is. Here, complexity is based on how many departments the project involves, because each department has its way of doing things, and it might be inflexible in accepting changes.

Part IV describes politics at the management level and at the project level. Management of politics at the management level is important, and it must be managed at the front end. The politics can be divided in three categories: 1) organizational issues, 2)

leadership issues, and 3) project management issues.
Management of politics at the project level is also
challenging because of the behavioral dimensions of
stakeholders, and it must be done throughout the
project life cycle.

Politics at the project level can be divided in
three categories: 1) stakeholders' management
issues, 2) project management and team leadership
issues, and 3) political issues related to managing
upward. Part IV describes politics associated with
all these issues and outlines several practical techniques to deal with politics at both levels and resolve
issues related to all categories.

The Art of Positive Politics

Three Truths of Life to Manage Stakeholders

Simplicity is the ultimate sophistication.
—Leonardo da Vinci

Politics are ways to get things done in an organization. Some find them unpleasant and distasteful, but they are an undeniably important and prime force. Despite good planning, many projects are derailed because of politics. Therefore, project managers must understand and manage politics to deliver successful projects. People skills are crucial to manage politics. Project managers must remember the three truths of life I learned from my parents, the best mentors of my life, when dealing with stakeholders and managing politics effectively. These three truths are shown in figure 9.1.

Three Truths of Life
1. People make or break things
2. People do mostly what is in their best interest —Everyone is looking for "What's In It For Me" (WIIFM)
3. People support what they create
Politics are remembering the above three truths and then managing your stakeholders accordingly.

Figure 9.1. Three truths of life.

9.1. Truth #1: People Make or Break Things

> *I will pay more for the ability to deal with people than any other ability under the sun.*
> —John D. Rockefeller

People make things happen, and people prevent things from happening. Wherever there are people, there are politics. People behave differently at an individual level, team level, or management level.

9.1.1. Individual level

> *People can alter their lives by altering their attitudes.*
> —William James

Unlike computers and machines, people have feelings. People behave in a certain way at an individual level. They have their way to do things and expect others to agree with them. They feel happy when you deal with them positively, and they feel sad when you hurt their feelings. The following are two rules to deal with people individually:

1. Deal with others the way you want to be treated — You must respect everyone for his or her feelings and ideas. Just as you do not want to be belittled or criticized for trivial things, you must treat others in a way they feel appreciated and respected. You must separate the person from his or her ideas; evaluate the ideas, not the individual.

2. Seek first to understand, and then to be understood — Stephen R. Covey (2000) describes this habit in *Seven Habits of Highly Effective People.* [1] This habit emphasizes that you must be empathetic and see things from the other person's perspective, which is crucial for effective communication and negotiations. Many people push their ideas on others. However, the effective way is to first understand the other person in the following contexts:

    a. What are their viewpoints and why?

    b. What are their feelings and beliefs and how strong are those feelings and beliefs?

    c. What are their interests and positions?

You must ask relevant questions to clarify points to understand the other person before pushing your ideas.

Most people don't care how much you know, but they appreciate how much you care about them. You must develop good people skills to communicate, motivate, and negotiate effectively. People are motivated to put in extra effort when they are happy, feel respected, and like their work environment.

9.1.2. Team level

Through teamwork, ordinary people can produce extraordinary results. They can lift things that come into their hands a little higher, a little further on towards the Heights of excellence.

—Henry Ford

Projects are done by teams composed of a mixture of people with different backgrounds, interests, and expectations. They must work together effectively to achieve common goals. Project managers must create synergy among their teams in a project environment where they are continually challenged to do with fewer resources. Project managers must develop good interpersonal skills and leadership skills to gain commitment from their stakeholders. They should create high-performance teams by practicing the following four guidelines:

1. Support your teams — You must support all team members. Project managers should follow the rule: When taking credit, give it to your team; when taking blame, take it on

yourself. Individual team members might make a few mistakes, but you must allow them to realize their mistakes and correct them. Give people opportunities to learn from their mistakes, which is effective if all team members are convinced of objective support from their managers.

2. Provide challenging opportunities — People like to expand their potential and want to learn new things, and they are often prepared to take challenging opportunities to do different tasks. Mentorship and coaching from project managers to meet new challenges increases team members' competency. Project managers can get extraordinary results from ordinary people by providing them challenging opportunities and reinforcing them positively.

3. Demonstrate trust and confidence in them — Many project managers hesitate to assign difficult tasks to some team members. They consider it risky because they are accountable for those tasks and overall team output. Consequently, some project managers micromanage those team members assigned the difficult tasks, which is even more common when the tasks are critical to the project's success.

4. Nurture creativity and innovation — Innovation is a key to organizational success. Project managers must create an environment where people are encouraged to think outside the box and feel comfortable in taking new challenges. They should act as a coach

317

to nurture creativity and innovation among team members. This approach develops ideas for new products and services and encourages team members to challenge the status quo and develop creative plans to improve existing processes and increase the project team's overall performance.

Teamwork is crucial to project success. Project managers must understand team dynamics and the importance of creating synergy among all team members. They must develop and use appropriate team-building skills to optimize the output of all team members individually. They must then create an environment where everyone works together and commits to helping one another increase the overall team performance.

9.1.3. Management level

Hold yourself responsible for a higher standard than anybody else expects of you. Never excuse yourself.
—Henry Ward Beecher

Management level refers to managing upward to get the required project support. As a first step, project managers should develop a good champion to build support for their project. Use the following guidelines to manage upward:

- Follow management hierarchy. (Do not go over your boss' head!)
- Make your boss look good.

- Align project goals to organizational goals to secure senior management's support.

- Focus on organizational goals and strategies and do everything to make things happen to meet organizational strategies.

Some think of managing up as political game playing. Instead, it is a positive way to work with your supervisors toward mutually important goals best for the organization. By managing up, project managers can build a productive working relationship with their bosses and get the resources needed to deliver successful projects (Carlone & Hill 2008, v–vii, 4–7).[2] Politically sensible project managers understand that they and their bosses depend on one another. Deborah Singer Dobson suggested that managing up should be considered constant, "like being in a good marriage." She suggests that once a trusting, mutually beneficial relationship is established, nurturing and evaluating it long term is easier (Dobson & Singer 2000).[3] Project managers can increase their effectiveness by establishing a stronger relationship with their bosses. They should use the following guidelines to manage upward successfully:

- Develop a strong positive relationship by understanding yourself and your bosses' expectations and promoting their goals.

- Communicate effectively with your bosses by listening to them actively, understanding and adapting to their communication styles, and handling disagreements with your bosses by constructive face-to-face discussions.

- Negotiate with your managers by demonstrating your credibility, identifying priorities, and presenting possible solutions, and not just problems. Negotiate for win-win solutions.

People are the backbone of most organizations. Project managers must understand that people make things happen or become roadblocks. You must develop good people skills to manage people at an individual level, team level, or at the upper management level. You must use good team management skills to create high-performance teams.

Managing upward is challenging, but required, to gain support from top management to get the resources required to meet project objectives. Project managers should develop a good working relationship with their senior managers by aligning project goals with organizational goals. They should develop creative solutions for project problems and make their bosses look good by delivering successful projects important to their bosses and their future.

People are the backbone of most organizations. Project managers must understand that people make things happen or become roadblocks. Project managers must develop good people skills to manage people at an individual level, team level, or upper management level. They must use good team management skills to create high-performance teams.

Managing upward is challenging, but it is required to gain support from top management to get resources required to meet project objectives. Project managers should develop a good working relationship with their senior managers by aligning project goals with organizational goals. They should develop creative solutions for project problems and make their bosses look good by delivering successful projects important to their bosses and their future.

9.2. Truth #2: People Do Mostly What Is in Their Best Interest

> *Successful people are always looking for opportunities to help others. Unsuccessful people are always asking, What's in It for Me?*
> —Brian Tracy

Most people ask What's in It for Me (WIIFM) before they help or give resources to anyone. You must recognize this truth because the WIIFM concept determines how people respond to your requests to work on your projects and initiatives.

Politically naive project managers assume that because they have prepared a sound plan for their proposed initiative or project in schedule, budget, and risk analysis, their project is good for the organization. Therefore, management and other stakeholders should support it. They expect functional managers to provide required resources with a proper skill mixture to meet project objectives. These project managers must understand that a new project's beauty is in the eyes of the beholders. You might agree with a project's benefits, but you must also convince other managers to get funding and people to execute the project successfully. This convincing process can be challenging, and it requires understanding of the following points:

- How do you identify people whose support is important?

- Why should others support this initiative?

- How can you help them advocate for your project?

- How could others meet their interests by supporting this initiative?

Project managers should address these questions and prepare answers and analysis in terms of What's in It for Them (WIIFT). They should look at the project outcomes from the perspective of others from whom they need support. You must invest appropriate efforts to understand and use politics to cultivate a good relationship with power holders to get their support. The process to gain others' support is similar to finding a good champion for the project. Politically sensible project managers prepare the answers for these questions as follows:

- Approach people in the organization who have high positional power and enough influence to help meet project goals and who benefit most from the proposed project's outcomes.

- Align project benefits to organizational objectives and department objectives.

- Provide good data, facts, and analysis for champions to help them advocate for your project successfully.

- Explain project benefits from champions' perspective. In other words, address What's in It

for Them (WIIFT) by speaking their language
and explaining benefits in terms of what is
important and relevant to them.

Politically sensible project managers ensure at
the front end that they approach the right people
to champion their project and secure the necessary
support. They recognize that champions should have
sufficient organizational influence to make things
happen in their favor and continually push their
projects ahead. These champions should have an
appropriate mixture of knowledge, influencing skills,
and network power to maintain high priority for the
project. They also should have the right positional
power (legitimate power) they can use, if needed, to
push the project by getting the required resources.

9.3. Truth #3: People Support What They Create

> *If everyone is moving forward together,*
> *success takes care of itself.*
> —Anonymous

Logically, when people are involved in something, they support it. Involving people at the front
end of creating projects or initiatives also makes
them feel important. They feel integral to the project
and *want* to do their best, rather than *have* to do
their best. Involvement or genuine participation
leads to their agreement and acceptance, which
leads to their support and buy-in, as shown in figure
9.2.

Figure 9.2. Gaining support through genuine involvement.

For example, during project planning, if stakeholders are genuinely involved in preparing the project plan, including the project budget, schedule, and risk analysis, they likely agree with and accept the plan with assumptions and constraints. Their participation and involvement must be genuine, and not lip service. If people have questions and concerns about the project plan, you must address their comments and feedback and evaluate them immediately, incorporating as much as possible in preparing the final project plan. In such cases, the involved stakeholders feel positive and support the project plan because they were part of its creation. These stakeholders are committed to make the plan work despite various challenges and problems.

If they were not involved in preparing the project plan, they lack a sense of ownership, and they are uncommitted to meeting the plan. You must involve most stakeholders genuinely in establishing new processes and methodology to implement them successfully.

Project managers must recognize the following points to gain their stakeholders' support:

- Participation should be genuine, and people should give their comments.

- People should be thanked for their gift of feedback.

- Agreements about ideas must be reached by convincing, rather than imposing. The following three conditions must be met to ensure that the acceptance reached and the agreements made for initiatives are implemented:

  1. Agreements must be free of threat, fear, or coercion.

  2. Agreements must be fair and reasonable.

  3. Good mutual understanding must exist between the parties.

If the three conditions are unmet, stakeholders will unlikely support the agreements reached and will get out of them however they can. Project managers must establish good working relationships with stakeholders and reach a *meeting of the minds*, meaning that if the relationship is solid and sincere with good mutual understanding, both parties will do their best to implement the agreement and commit to meeting objectives.

People naturally support what they create. So, it is important to involve people genuinely to seek their input and feedback for our proposed projects/

initiatives and then incorporate their comments as much as possible in creating the final plans and processes to meet objectives. Any feedback received should be considered as a valuable gift and people must be thanked and appreciated for this, which makes people feel important and appreciated and likely to support your initiatives.

Chapter 9 Summary

The three truths of life described in this chapter highlight the importance of understanding people and their behaviors to get desired results. It is sensible to understand how people behave at in interpersonal level as well as at the team level and how their behavior and actions can be influenced to deliver successful projects. In general, when people are appreciated and treated with respect, they are motivated and committed to produce better results.

When people are asked to help, they often ask the question, What's in It for Me (WIIFM), that is, why should they help us? Therefore, you should approach people and departments for help by addressing the question, What's in It for Them (WIIFT). This forces you to devise a strategy to influence them and gain their support. People naturally support what they create. So, you should involve people genuinely to seek their input and feedback for your proposed projects/initiatives and then incorporate their comments as much as possible in creating the final plans and processes to meet project objectives. Any feedback received should be considered a valuable gift, and people must be thanked for it, which makes people feel important and appreciated, and they are likely to support your initiatives.

People are the backbone of most organizations. Project managers must understand that people make things happen or become roadblocks. Project managers must develop good people skills to manage people at an individual level, team level, or upper management level. They must use good team management skills to create high-performance teams.

Managing upward is challenging, but it is required to gain support from top management to get resources required to meet project objectives. Project managers should develop a good working relationship with their senior managers by aligning project goals with organizational goals. They should develop creative solutions for project problems and make their bosses look good by delivering successful projects important to their bosses and their future.

At the front end, politically sensible project managers ensure that they approach the right people to champion their project and secure the necessary support. They recognize that champions should have sufficient organizational influence to make things happen in their favor and continually push their projects ahead. These champions should have an appropriate mixture of knowledge, influencing skills, and network power to maintain high priority for the project. They also should have the right positional power (legitimate power) they can use, if needed, to push the project by getting the required resources.

Managing Politics at the Management Level

*To succeed in politics, it is often necessary
to rise above your principles.*

—Anonymous

Politics are inevitable in most organizations and especially in project environments because projects are done by team members with different values, beliefs, interests, and expectations. Most organizations have limited resources; therefore, project managers compete to get resources with the appropriate skill mixture for their projects by influencing senior management and resource managers. Others perceive this approach of "getting things done by going around the normal channels" as politics.

There are politics at all organizational levels. Verma (2012–2015; 1996, 239–242) described excellent guidelines for managing politics at the upper management level and at the project level.[1] Most people wonder about the following questions and issues related to politics:

- Why do we have politics?
- Can organizations be free of politics?
- Who engages in politics and why?
- How do we become politically sensible?
- How can we minimize the negative impact of politics?

Projects are done by people with different interests and ideas about how things should be done. Wherever there are people, there are politics. In addition, most organizations lack infinite resources. Often, conflict and competition arise over limited resources. Other issues that might lead to disagreements and conflicts include allocation of power and responsibilities, bonuses, and budgets. Therefore, most organizations will unlikely ever be free from politics because of such conflicts.

Two types of people engage in politics: 1) people who use politics to meet their objectives, even at the cost of others (Sharks) and 2) Politically sensible people (Dolphins) who try to meet organizational objectives and think of win-win solutions. Project managers must learn to become politically sensible by aligning project goals with those of the organization.

Many people think senior management is responsible, either for creating politics themselves or letting members of senior management get away with playing politics. Therefore, you must understand what senior management can do to minimize the negative impact of politics.

Senior management can play a major role in minimizing the negative impact of politics. One approach is to discourage negative politics and encourage the concept of positive politics. Senior management should understand both negative and positive types of politics and the important keywords for each type. This understanding helps redirect people's energy from the negative side of politics that includes blaming and manipulating others to the positive side that emphasizes creating a culture of high integrity, collaboration, synergy, and open communication.

Managers at all levels should be encouraged to allocate resources and share information among stakeholders objectively. Senior management should look into how the organization operates compared with how it should operate. They should pay special attention to three types of areas or issues to minimize the negative impact of politics.

10.1. Organizational Issues

Success is the result of good judgment; good judgment is the result of experience; and experience is often the result of bad judgment.
—Anthony Robbins

Senior management is responsible to manage the organization. Organizational issues deal with how the organization is structured and how it operates, covering high-level policy and procedures to address issues related to reporting relationships, communication channels, performance appraisals and reward systems, mentoring programs, and special problems.

Table 10.1 shows guidelines to minimize the negative impact of politics related to organizational issues.

Issues	Guidelines & Tips
Organizational Issues	• Develop a clear organizational structure and communication system. • Implement a management by projects (MBP) approach across the organization. • Create objective reward systems tailored to organizational goals. • Develop performance goals with an appropriate plan. • Engage external consultants as SMEs, lobbyists, and agents. • Develop and implement a mentorship program.

Table 10.1. Managing Politics at the Management Level (Organizational Issues)

10.1.1. Develop a clear organizational structure and communication system

> *Skill in the art of communication is crucial to a leader's success. He can accomplish nothing unless he can communicate effectively.*
>
> —Anonymous

Unclear organizational structures and communication systems cause many problems. Such problems might be related to operational issues, communication breakdowns, and conflicts. Project managers and project personnel assume how the organization should run and how communication lines should flow. These assumptions might be wrong, leading to serious problems in meeting project objectives, with the negative politics of finger-pointing and unnecessary blaming. Senior management can solve this problem from the start by developing a clear organizational structure, chains of command, and a clear communication system to encourage collaboration, sharing of information, and teamwork.

10.1.2. Implement a Management by Projects (MBP) approach across the organization

Management by objectives works if you first think through your objectives. Ninety percent of the time, you haven't.

—Peter Drucker

Today's business world faces serious challenges with international competition, resource limitations, economic constraints, downsizing, and changing technology. As markets become more dynamic and global, MBP is being recognized as one of the most effective ways to manage an organization. Without this approach, people might work on their favorite initiatives and use resources that should be allocated to high-priority projects.

Senior management must develop organizational strategies and then design effective programs and project management systems to meet organizational goals within time and budget constraints. Senior management must implement this approach carefully across the organization by understanding organizational culture, engaging people at all levels, listening to their feedback, and incorporating their comments, as necessary, to prepare long-term and short-term plans. Program and project managers should be given proper guidance and support to deliver successful projects. All human resources must be accounted for in their contribution to selected projects. However, innovation is the key to long-term success, and therefore, appropriate resources must be allocated to research and development projects.

10.1.3. Create objective reward systems tailored to organizational goals

> *If you pick the right people and give them the opportunity to spread their wings—and put compensation as a carrier behind it—you almost don't have to manage them.*
> — Jack Welch

Unfair and subjective reward systems often cause negative politics. Sometimes, people are rewarded because of favoritism and their relationships with senior managers. This behavior often frustrates many people who believe they are improperly recognized for their performance. Consequently, they lack motivation, and they are likely to engage in negative politics.

Senior management can minimize the negative impact of politics by developing and implementing clear and objective criteria for distributing rewards and recognition. In consultation with managers, they should develop a clear performance evaluation system that discourages subjectivity. This approach increases employee motivation and commitment, leading to higher performance to meet organizational strategies and goals. Senior management must consider the following factors while designing

a reward system and associated policies and
procedures:

- Rewards must be based on objective criteria.

- Rewards must satisfy the need people value
  the most.

- Rewards must be tied to contributions to
  overall organizational performance.

Senior management should minimize frustration
and negative politics unfair reward systems cause,
by eliminating subjectivity and favoritism and understanding what rewards turn on people and help
them be motivated long term.

10.1.4. Develop individual performance goals with an appropriate plan

> *People are more inclined to be drawn in if*
> *their leader has a compelling vision. Great*
> *leaders help people get in touch with their*
> *own aspirations and then will help them forge*
> *those aspirations into a personal vision.*
> —John Kotter

People determine any organization's success or
failure. They plan, organize, execute, and monitor
programs and projects. They use combined project
management and interpersonal skills to meet organizational goals. You can view people as problems
and constraints or as solutions and opportunities.
Senior management must focus on developing
people and helping them grow. They should create

an environment where people feel motivated, supported, and committed to increase their personal performance as well as team performance to meet organizational goals and strategies.

Everyone's performance is possibly lower than senior managers like. Senior management should develop a clear performance evaluation system based on objectivity and sincerity. The main purpose of the performance evaluation systems should be to identify problem areas with a focus on how to improve performance. Many managers are not well trained in evaluating performance objectively with constructive feedback.

The lack of clear, objective, and sincere performance evaluation systems leads to anxiety, low confidence, poor morale, and lack of motivation among employees. It increases the likelihood of people blaming others, becoming defensive, and getting involved in negative politics. Senior managers should be encouraged to work with project managers to develop project team members' individual performance goals with appropriate plans. Such plans involve training, coaching, and providing new opportunities and challenges. Senior management must continually provide proper resources, guidance, and support that achieve extraordinary performance and minimize the negative impacts of politics.

10.1.5. Engage external consultants as subject matter experts, lobbyists, and agents

No one can whistle a symphony.
It takes a whole orchestra.
—H. E. Luccock

Senior management is responsible to provide leadership to run the organization effectively. Project managers expect senior management to be knowledgeable and expert in many areas related to project issues. They expect senior management to understand in-depth analysis of project issues and provide proper guidance to resolve them.

However, for project issues highly specialized in subject matter, technology, policies and procedures, and politics, such as with global projects, defense organizations, and pharmaceutical industries, senior managers might lack the necessary knowledge to understand and carry out an in-depth analysis of the situation. In such cases, after consulting with project managers, senior management should engage consultants with a good reputation and special knowledge to help them understand the situation and get their advice for subsequent decisions and actions.

Senior management can engage outside consultants as subject matter experts, lobbyists, and agents to seek their advice.

Senior management likes to do this to get an independent and neutral opinion, but project managers might view it as politics to delay decisions and actions. Most project managers feel positive about doing so because it provides them an opportunity to learn more about the situation and issues. Politically, they might also feel relieved from backlash, should their analysis and recommendations be followed and later proved wrong.

10.1.6. Develop and implement a mentorship program

Tell me and I forget; teach me and I may remember; involve me and I learn.

—Benjamin Franklin

Every organization has a unique culture and certain ways of doing things. People with long working experience in such organizations have learned techniques (tricks of the trade) to deliver successful results. New project managers might not know the so-called *tricks of the trade* to use the processes and deal with people effectively. They might have to work much harder and still face difficulties and problems in getting things done to meet project objectives. Sometimes, they might become frustrated when they see others getting things done with much less effort.

Senior management can minimize frustration and anxiety among younger people by developing and implementing a mentorship program led by the human resources department or by the leader of a large project. In this program, you can assign experienced program and project managers with good people skills as mentors to help younger people increase their knowledge and experience in managing people and projects. These mentorship programs must be well planned and supported by senior management with proper guidance, rather than lip service.

Most organizations face the challenges of economic uncertainties, global competition, resource limitations, and a lack of highly motivated people. Senior management is responsible to provide the right leadership to achieve the highest level of customer satisfaction for their products and services as well as satisfied and highly committed employees. A lack of proper organizational leadership creates conflicts for scarce resources, poor teamwork, and obstacles for people in doing their jobs. It creates stress and frustration leading to negative politics.

Senior management can take many actions to minimize the negative impact of politics. They must develop clear organizational structures and communication systems to operate the whole organization efficiently. They should implement MBP across the organization to optimize the use of resources. Performance evaluation and reward systems should be tailored to organizational goals and be based on objective criteria. Senior management should engage subject matter experts to get independent advice about special problems and outside consultants to act as lobbyists, arbitrators, and agents for special issues to help promote the business in large government departments and pharmaceutical departments. Finally, senior management should develop and implement a mentorship program.

10.2. Leadership Issues

*Leadership is the capacity to
translate vision into reality.*
—Warren Bennis

Senior management is expected to provide good leadership. Leadership issues focus on providing long-term vision, realistic goals, priorities, guidance and support, and clear directions, which involves creating an environment where people feel inspired, motivated, empowered, and committed and a culture of high collaboration, synergy, and teamwork to produce extraordinary results. Leaders should have high integrity and be willing to improve their leadership skills and those of other managers.

Guidelines to minimize the negative impact of politics related to leadership issues are shown in table 10.2.

Issues	Guidelines & Tips
Leadership Issues	• Lead by example and have zero tolerance for negative politics. • Develop clear vision and objectives for the organization and departments. • Hire senior managers with high EQ and matching organizational values. • Empower project managers and distribute power to avoid power corruption. • Assign champions for high-priority and high-profile projects. • Provide 360-degree feedback.

Table 10.2. Managing Politics at the Management Level (Leadership Issues)

10.2.1. Lead by example and have zero tolerance for negative politics

> *If we want the men to do a good job for*
> *us, we must do a good job for them.*
> —John F. Kennedy

Leaders should have high integrity, sincerity, and personal conviction to meet organizational strategies and goals. Instead of just expecting others to work hard, they should be committed to work tirelessly and help everyone in the organization expand his or her potential to deliver successful programs and projects. Project personnel often feel frustrated and demotivated when they discover senior managers engage in negative politics to get things done, and they cannot trust their leadership skills to make timely decisions and take proper action.

Senior management should lead by example and walk the talk by communicating effectively, thinking of win-win negotiations, supporting team members, coaching, and mentoring, as necessary. They should understand the dynamics of politics and their negative effect on organizational performance. They should know many political behaviors that lead to negative politics and then manage those behaviors effectively. They should have zero tolerance for politics (especially negative politics) and nurture a climate of collaboration, synergy, and teamwork. Often, when organizational employees see high integrity, openness, and personal commitment from senior management, they commit to work toward organizational goals, rather than playing negative politics to meet their interests.

10.2.2. Develop clear vision and objectives for the organization and departments

A goal without a plan is just a wish.

—*Larry Elder*

Project managers expect their senior managers to have clear vision and select a portfolio of programs and projects to meet overall organizational strategies for the organization and major departments. Significant ambiguity in the systems for selection of programs, prioritization, and resource allocation causes negative politics. It creates confusion, frustration, and anxiety among project managers. They assume and then act accordingly because of the lack of proper directions and timely decisions from senior management, possibly wasting resources, delivering unsuccessful projects, and creating negative politics. Then, people blame one another and find scapegoats.

Senior management can resolve this problem and minimize the negative impact of politics at the front end by developing clear and effective systems to select programs, set priorities, and decide about allocating scarce resources. They should ensure that selected programs and projects align with overall organizational strategies as well as the goals of all major departments and business units, curtailing confusion, lack of trust, and working at cross-purposes.

10.2.3. Hire senior managers with high emotional quotient (EQ) and matching organizational values

When dealing with people, remember you are not dealing with creatures of logic, but with creatures of emotion.

—Dale Carnegie

Senior management leadership quality is the key to organizational success. You must hire all managers with good leadership skills and high moral values, especially senior managers, because they are expected to set examples for establishing and following high organizational values. Senior management should have excellent leadership skills in addition to managerial skills. They must direct organizational resources toward doing the right things at the right time to meet constraints of global competition, economic uncertainties, and limited resources. Good leaders should have high emotional quotient (EQ) and inspire and motivate all employees to meet organizational goals. Therefore, when hiring senior managers, the following factors are considered due diligence:

- Look for high EQ in addition to high IQ.
- Ensure their personalities fit the organizational culture.
- Ensure they have high moral values.

Senior managers are expected to lead business units or departments and finding the best candidates for senior management is challenging. Finding the

best senior managers with high moral values is
sometimes difficult, but great efforts must be made
before making final decisions. These efforts include
attitudinal and situational questions, reference
checks, and information from informal networks, for
example, associations, clubs, and so on.

10.2.4. Empower project managers and distribute power to avoid power corruption

> *Leaders should influence others in such a*
> *way that it builds people up, encourages and*
> *educates them so they can duplicate*
> *this attitude in others.*
> —Bob Goshen

Senior managers have combined formal power
and informal power. Some managers are granted
formal power based on their position in the organizational hierarchy, which gives them freedom to
allocate resources (financial and people), make decisions, hire or fire employees, and distribute rewards
and punishments. Project managers must have great
informal power, which they must earn with their
knowledge, experience, expertise, and leadership
skills.

Besides having power, they must know how to
use it effectively to get things done successfully
through people. Project managers prove effective
project managers and good leaders if they focus on
increasing their informal power. As a strategy to use
their power, they should develop skills to use their
overall power (formal and informal) informally most
of the time and use their power formally only as a
last resort.

Senior management must understand the dynamics of power and know how power and politics are interrelated. The misuse of power can increase stress, anxiety, resentment, and poor morale, leading to negative politics. Senior management can use the following guidelines in distributing power in an organization to minimize the negative impact of politics:

- Develop flatter organization structures to distribute power across more functional departments, because if power resides in the hands of only a few managers, there is danger of power corruption.

- Empower project managers to increase their sense of ownership and their ability to accept new responsibilities and challenges and manage their projects effectively to deliver successful results.

Empowerment is a different way of working together. *Empowerment* means giving managers more freedom over their jobs in a way they experience a sense of ownership. It involves giving them more authority to make decisions, more resources, and more information to help them do their job more effectively and requiring accountability to achieve higher commitment, better results, and responsible use of power. Empowerment influences employees, teams, and organizations in how they work, feel, and achieve results (Wellins, Byham & Wilson 1991, 21–23; Scott & Jaffe 1991, i–ii, 4–16, 29–37).[2]

Empowering project managers increases their skills and confidence as well as those of their team

members. Limiting power centralization among a
few departments or managers also can help avoid
the risks of power corruption. Senior management
can use these techniques to minimize the negative
impact of politics, which typically leads to high
stress, poor morale, and low performance.

10.2.5. Assign champions for high-priority and high-profile projects

> *Good leaders know when to display*
> *emotions and when to delay them.*
> —John Maxwell

Projects are done by people with different interests, viewpoints, expectations, and backgrounds. In
organizations, a few people might be self-centered
and try to meet their objectives and engage in negative politics. They might create obstacles for project
managers in delivering successful results. Having
the right champions is the key to delivering successful programs and projects. Project managers depend
a great deal on their champions' effectiveness. For
project managers, good champions are expected to
fulfill two key roles:

1. Champions are interested in your projects,
   and they want you to succeed.

2. Champions use their power and influence to
   remove your obstacles.

Typically, project sponsors are perceived as
project champions as well. However, project sponsors
might or might not be good champions in fulfilling

the two roles. In addition, champions must have appropriate positional power, depending on project complexity, to help project managers manage their projects effectively. Different departments and business units have their way of doing things, and therefore, there are more politics when a project uses people from different departments. Here, the number of different organizational departments, business units, and functional disciplines involved in the project defines complexity.

Most project managers face many difficulties and the negative impact of politics in managing their projects if they lack a good champion with the right positional power and the ability to influence stakeholders effectively. Senior management can help project managers in this context by helping them find a champion and assigning senior managers at the right organizational level as champions for high-priority projects.

10.2.6. Obtain 360-degree feedback

> *Great people talk about ideas, average*
> *people talk about themselves, and*
> *small people talk about others.*
> —John Maxwell

Senior management is responsible to provide good leadership continuously for their organizations. In today's business environment, projects involve people from different professional backgrounds, cultures, generations, and lifestyles. Senior management can turn off people if they do not take time to understand project personnel's characteristics and

associated strengths and weaknesses, resulting in
resentment, lack of cooperation and commitment,
and lack of confidence in senior management's
leadership skills. All these factors might lead to poor
morale and negative politics. Therefore, leadership
styles must match organizational culture, personalities, and professional strength of personnel and
generation (age) of the workforce.

Today's workforce expects senior management
to use a participative and collaborative leadership
style, rather than autocratic or authoritarian style.
Senior managers should be open-minded and open
to suggestions and feedback from their employees to
improve their leadership styles and be more effective
in producing better results. They should have a
system of 360° feedback to regularly evaluate their
leadership skills to determine the following:

- How staff members rank senior management's leadership skills

- Major issues and problems

- Strategies to improve and develop effective
  leadership skills

There are many diagnostic instruments to
evaluate leadership skills. Kouzes and Posner
(2007, 3–27) developed five practices of exemplary
leadership and an associated Leadership Practices
Inventory (LPI) tool to evaluate the leadership
practices of leaders and team members with recommendations to improve leadership skills to deliver
successful results.[3] When using any evaluation
system, you must receive feedback as a gift and be

open-minded to consider suggestions to improve leadership skills seriously.

Politics are inevitable in business organizations and especially in project environments. Senior management is expected to provide high-quality leadership to meet organizational strategies by delivering successful programs and projects. Lack of good leadership can create high stress, poor morale, and lack of commitment. Senior management can minimize the impact of these undesirable factors and negative politics by using several guidelines in addressing leadership issues: leading by example, hiring senior managers with high EQ and moral values, and demonstrating zero-tolerance for negative politics.

Senior management should develop clear vision and organizational strategies for the organization and various departments, distribute power to avoid power corruption, empower project managers and project team members to increase their sense of ownership, help project managers find good champions and allocate champions with an appropriate power level in the organization for high-priority projects. Senior management should be open-minded and prepared to adapt their leadership skills and styles to match employee maturity, and use 360°–feedback to determine their leadership effectiveness and receive suggestions and feedback from their employees to improve their leadership skills and styles continually.

The Art of Positive Politics

10.3. Project Management Issues

> *In poorly run projects, problems can go*
> *undetected until the project fails. It is like*
> *the drip, drip, drip of a leaking underground*
> *pipe. Money is being lost, but you don't see it*
> *until there is an explosion.*
>
> —Joy Gumz

Senior management must ensure programs and projects are properly selected and managed. Project management issues deal with providing clear directions about the prioritization system and resource allocation. Senior management must provide enough financial and organizational support to project managers continually to deliver successful programs and projects. They should develop effective project management systems and methodologies in collaboration with project managers and stakeholders.

Guidelines to minimize the negative impact of politics related to project management issues are shown in table 10.3.

Issues	Guidelines & Tips
Project Management Issues	· Develop and follow a clear process for project selection. · Clarify roles and responsibilities for project sponsors and project managers. · Create an evaluation system to analyze and approve projects. · Provide training, as needed, to improve project management and people skills. · Avoid stopping and restarting a project, as it reduces momentum. · Enforce an amnesty rule (no punishment for providing honest data).

Table 10.3. Managing Politics at the Management Level (Project Management Issues)

10.3.1. Develop and follow a clear process for project selection

A business plan is like a war plan;
when your competitors know about it,
it's no longer of any use.
—Bangambiki Habyarimana

Senior management is responsible to provide proper directions for project managers, project teams, and all project stakeholders to ensure everyone works together toward a common purpose. Lack of clear directions creates confusion among project personnel. Consequently, they start assuming what should be done in the organization, when, and who should do it and how. The differences in assumptions and associated actions might waste resources and

351

create negative politics. The following are main issues or questions senior management must address
to provide clear long-term and short-term directions
and create synergy:

1. How are the projects selected and why?

2. What is the system for setting project
   priorities?

3. How are decisions made about resource
   allocation?

To answer these questions, senior management
must develop and implement clear systems and
processes for project selection and prioritization and
resource allocation to avoid confusion and negative
politics. Most organizations can only work on limited
programs and initiatives because they have finite
resources. Several business units and departments
might suggest programs, and senior management
should evaluate them to select the best.

Various stakeholders try to push their programs
and lobby to get higher priority. However, senior
management should develop systems based on objective criteria and long-term stakeholder satisfaction.
To optimize the system for selecting and prioritizing
programs and initiatives, senior management must
first develop overall organizational strategies and
then use an objective evaluation system to select
projects that align with organizational strategies.
Similarly, various projects' priorities and resource
allocation should be decided based on how closely
they align with organizational goals and strategies.

The evaluation process should be designed
to seek input from major internal and external
stakeholders. This evaluation process is transparent
for research facilities, educational institutions, and
publicly funded organizations. However, for some
businesses and government departments (such as
the Department of Defense, Homeland Security,
the National Energy Board (NEB), the Canadian
Nuclear Safety Commission (CNSC)), it might be
classified for security and business confidential-
ity. For example, large international world-class
research facilities select programs and initiatives
and decide priorities based on international motivation and collaboration, importance in scientific and
research objectives, overall budget and timeline,
availability of technology, specialized resources, and
so on.

10.3.2. Clarify roles and responsibilities for project sponsors and project managers

*Talent is the multiplier. The more energy and
attention you invest in it, the greater the yield.
The time you spend with your best is, quite
simply, your most productive time.*
—Marcus Buckingham

Clear definition of roles and responsibilities is
the key to successful projects. Lack of clarity in these
areas leads to misunderstandings, communication
problems, and risks of important things falling
through the cracks. People become over defensive,
blame one another, and push their agendas, leading
to negative politics. Senior management should understand and analyze such issues and develop clear

roles and responsibilities for all project stakeholders, especially the following:

- Project sponsors for overseeing projects and providing timely directions and decisions
- Project champions for supporting project managers and removing roadblocks
- Project managers to manage project activities throughout the project life cycle by using their technical and interpersonal skills
- Project team members and other stakeholders for working together and delivering results with emphasis on quality
- Functional managers to provide resources with appropriate skill mixture, as needed to meet project objectives

Clear definition of roles and responsibilities for these positions and for other project personnel avoids misunderstandings and creates a sense of ownership and teamwork. Senior management can minimize the negative impact of politics and create a culture of more collaboration and teamwork.

10.3.3. Create an evaluation system to analyze and approve projects

In this business, by the time you realize you're in trouble, it's too late to save yourself. Unless you're running scan all the time, you're gone.
—Bill Gates

Senior management can delegate responsibility to project managers for meeting their project objectives, but senior management is ultimately responsible to successfully complete an overall portfolio of programs and projects to meet overall organizational goals. Senior management should provide proper guidance and support to avoid panic and stress when problems are encountered in projects. Panic leads to inefficiencies, rework, and making wrong and irrational decisions while executing project activities. People become over defensive and play negative politics to deflect the blame to others. Efficient systems should be established to monitor projects continually to identify problems at early stages so corrective actions can be taken in time.

The project management system should implement the following tools to monitor projects effectively:

- Dashboard system to get the status of all programs and projects

- Gating process to ensure project phases are done in a proper sequence with quality

- Risk management system to identify and analyze risks and develop strategies to manage them

Senior management should develop systems to evaluate projects and receive honest and timely information to understand the problem areas, analyze options, and make final decisions to find the best solutions. They should make sure these systems are robust and efficient so the information and analysis is correct, avoiding panic and stress because of the lack of proper planning and execution of project activities.

10.3.4. Provide training, as needed, to improve project management and people skills

> *No great manager or leader ever fell from heaven; it's learned, not inherited.*
> —Tom Northrup

In project environments, people from various departments are allocated to a project on a matrix. They do their best to meet project objectives. However, in some cases, they lack the knowledge and skills to finish their assignments with high quality, but they might not express the need for training to their project managers. Training programs can be tailored to improve the following three types of skills of project personnel:

1. Technical skills associated with project management processes and the job's technical components

2. Interpersonal skills or soft skills to deal with people effectively

3. Leadership skills to inspire and motivate people to produce extraordinary results

Lack of training opportunities leads to lack of self-confidence, which leads to reduced performance and job satisfaction. Project managers must know their team members' strengths and weaknesses and develop strategies to capitalize on their strengths. Project managers should find an effective training program for their team members and seek required support from their managers. Senior management's genuine support in providing training as needed for

project personnel is perceived positively. Instead of
complaining about poor management and engaging
in negative politics, people feel more committed and
motivated to help management meet organizational
objectives.

10.3.5. Avoid stopping and restarting a project, as it reduces momentum

> *If you tell us early, it is our problem. If you*
> *tell us late, it is your problem.*
> —Vijay Verma

Senior management should develop good systems
based on objective criteria to properly select projects,
prioritize projects, and allocate resources because
a significant number of organizational resources
are allocated to selected programs and projects.
Sometimes, managers with great power and good
political skills push their favorite projects to meet
their interests. This wrong selection of projects leads
to wasted resources that could have been allocated
to do the correct projects. When such mistakes
are made, people lose confidence in management,
and they might engage in negative politics. Senior
management might detect this problem and try to
terminate favorite projects, but those managers
continue to use their political skills to push their
projects informally and use resources for the wrong
projects.

In such cases, senior management should call
out such negative political behaviors and terminate
those projects, starting a replacement project, if necessary. Senior management should have a funeral

ceremony for terminated projects to remind people associated with the projects that no more resources will be consumed for those projects. Such decisions can help, but starting and stopping projects reduces momentum and wastes organizational resources. Therefore, senior management should be careful to select the right projects from the beginning.

10.3.6. Enforce amnesty rule (no punishment for providing honest data)

> *In NASA, we never punish error. We only punish the concealment of error.*
> —Al Siepert

The amnesty rule implies that project managers are not punished for giving correct data and information about project status and problems even if they are unpleasant and undesirable. Senior management is responsible to monitor programs and projects.

Projects often encounter a few problems that cause cost overruns and schedule delays. Sometimes, project managers experience lack of support and constructive feedback from management to resolve project problems. Consequently, they fear telling the truth about project status to management early because of negative experiences, causing undue delay in discovering problems that might become too difficult to solve, even with more resources.

This perception of fear, threat, or coercion causes panic and stress when project problems are finally discovered, and they must be resolved quickly. Resistance, lack of cooperation, and poor performance leads to negative politics. Senior management can resolve this problem by establishing an amnesty rule where project managers feel free to present the true project status with potential risks and warning signals. With this approach, management learns the problems early enough to work with project managers and find proper solutions.

In this context, senior management should set rules and use a practice, while monitoring projects and discussing problems with project managers, that emphasizes the philosophy of "If you tell us early, it is our problem; if you tell us late, it is your problem."

The right programs and projects must be selected and managed effectively to meet organizational strategies and goals with limited resources. Senior management must resolve many problems related to project management issues that involve streamlining processes and systems to optimize organizational resources. Senior management should clarify the roles and responsibilities of project sponsors, project champions, and project managers to minimize any confusion and communication problems. They should establish efficient systems to monitor programs and projects continually and identify risks to find solutions promptly.

Senior management should create an environment to increase professional competence and personal commitment by supporting an adequate training program and encouraging project managers to provide honest and correct data and information to identify problems early enough to resolve them efficiently.

Chapter 10 Summary

Politics are often necessary to get things done, especially in project environments, because of competition and conflict for limited resources. Senior management must identify reasons for politics and develop strategies to minimize the negative impact of politics. This chapter describes three categories of issues related to managing politics at the senior management level—organizational issues, leadership issues, and project management issues—and presents various guidelines and tips to manage politics to resolve those issues.

Senior management is expected to develop effective policies and systems across the organization. They can deal with organizational issues that lead to negative politics by using the following guidelines:

- Develop a clear organizational structure and communication system.

- Implement an MBP approach across the organization to increase accountability.

- Create objective reward systems tailored to organizational goals.

- Develop individual performance goals with an appropriate plan.

- Engage external consultants as subject matter experts, lobbyists, and agents.

- Develop and implement a mentorship program.

Senior management is responsible to provide organizational leadership. To resolve politics related to leadership issues, they can use the following guidelines:

- Lead by example and have zero tolerance for negative politics.
- Develop clear vision and objectives for the organization and departments.
- Hire senior managers with high EQ and matching organizational values.
- Empower project managers and distribute power to avoid power corruption
- Assign champions for high-priority and high-profile projects.
- Obtain 360-degree feedback.

Senior management must meet organizational strategies and goals successfully by effectively managing portfolios of all programs and projects. Senior management can use the following guidelines to resolve project management issues:

- Develop and follow a clear process for project selection.
- Clarify roles and responsibilities for project sponsors and project managers.
- Create an evaluation system to analyze and approve projects.
- Provide training, as needed, to improve project management and people skills.
- Avoid stopping and restarting a project, as it reduces momentum.
- Enforce amnesty rule (no punishment for providing honest data).

Senior management might be unable to eliminate politics completely in an organization, but they must develop strategies and systems to minimize the negative impact of politics to deal with organizational issues, leadership issues, and project management issues by following the guidelines described in this chapter.

Managing Politics at the Project Level

*Stand up to your obstacles and do something
about them. You will find that they have half
the strength you think they have.*

—Norman Vincent Peale

Most project managers believe that because they interact closely with their team members and core stakeholders frequently, they understand their nature and viewpoints better. Therefore, they believe managing politics at the project level is a little easier than managing politics at the upper management level where they have to deal with senior management and stakeholders with higher positional power than themselves.

However, dealing with politics at the project level is challenging because of behavioral dimensions and because it is sometimes difficult to know the interests and concerns of team members and core stakeholders. The three categories of challenges or issues most project managers face while managing politics at the project level are the following:

1. Political Issues and Managing Upward
2. Project Management and Team Leadership Issues
3. Stakeholder Management Issues

11.1. Political Issues and Challenges of Managing Upward

You have to learn the rules of the game. And then, you have to play better than anyone else.

—Anonymous

Politics are inevitable in organizations and especially in project environments. Besides general management and project management skills, project managers must recognize the need to manage upward and deal with politics at the upper management level. They must learn to manage politics at all levels to meet their project objectives and keep stakeholders happy. Managing politics upward is critical, and it should be done at the front end or early enough in the project life cycle to avoid problems, disruptions, and roadblocks. Dealing with upper management and stakeholders higher than project managers in organizational hierarchy is challenging. Verma has described guidelines to deal with power and politics at the upper management levels to meet long-term organizational objectives, as shown in table 11.l (Verma 2012–2015, Verma 1996, 239–242).[1]

Issues	Guidelines & Tips
Political Issues & Managing Upward	• Follow management hierarchy. • Gain top management support by having good champions. • Have a good champion with appropriate positional power. • Develop a strategy to deal with special issues by engaging outside consultants. • Find out "What's In It For Them?" (WIIFT) to manage upward. • Analyze political landscape and protect yourself from sharks. • Learn to be politically sensible! (Relate project goals to organizational and departmental goals.

*Table 11.1. Managing Politics at the Project Level
(Political Issues and Managing Upward)*

11.1.1. Follow management hierarchy

*Organization charts and fancy titles
count for next to nothing.*
—Colin Powell

Project managers must follow management hierarchy when they escalate issues and deal with political situations. Circumventing a boss can get you in serious trouble, which happens often in progressive and flat organizations. Project managers must know their organization's internal culture and the importance of following the hierarchy in communicating various project management issues.

Not following management hierarchy in military, defense, and large bureaucratic organizations is political suicide. Although in most research, educational, and small organizations, people communicate and discuss issues more freely with many people from different disciplines and different levels, they should be careful when making decisions and dealing with sensitive personal issues.

11.1.2. Gain top management support by having good champions

> *Ensuring top management support*
> *is the key to project success.*
>
> —Vijay Verma

The best way to gain support from senior management is to have a champion who is enthusiastic about your project and willing to remove roadblocks for you when you encounter one, especially the political issues you can't resolve yourself. Gaining support from senior management and managing politics requires careful attention to the following two issues:

1. **Having a good champion** — The project champion is the most important advocate who removes obstacles and roadblocks the project manager faces. Project managers must first find a champion and then do everything possible to help their champion win battles for them. The following four questions help develop a process with practical tips for developing good champions:

   a. Where do you find a champion?

      Because champions should have sufficient organizational clout and influence to help the project manager and remove roadblocks, potential champions should be in upper management. They should belong to steering committees or other senior management committees (priority panels, advisory committees, and so on) where the following decisions are made:

- Project selection

- Priorities for all projects

- Resource allocation

b. How do you convince managers to be your champion?

Project managers can convince appropriate senior managers to be their champions by describing the project benefits and outcomes for their interests, that is, What's in It for Them (WIIFT).

c. How do you strengthen your champions? Project managers should try their best to strengthen their champions by giving them good data, information, facts, analysis, and proposed solutions. In a simple way, they should give their champion a few slides they can use to prepare and make presentations to senior management committees to advocate for resources and other help project managers require.

d. How do you sustain your champions? Project managers must thank their champions to sustain a good relationship with them. They should be ready to help their champions, as requested.

2. **Appropriate positional power for a champion** — Project managers must try to find champions from the appropriate level in the management hierarchy. This level depends on project complexity, which is defined by the number of organizational departments and disciplines involved. In addition, more politics are likely when more people from many departments are involved, because they come with different personalities and attitudes, new points, and belief in how things should be done. Different people push their agendas and try to influence others to follow their viewpoints.

A more complex project (that is, more departments are involved) requires a champion higher in the management hierarchy. For example, as shown in figure 11.1, project A's complexity is the highest followed by that of projects B and C. Therefore, project A's champion should be an executive VP, whereas a VP and director might suffice for projects B and C, respectively.

A director for project A is ineffective to remove roadblocks, as needed, because of the lack of organizational clout and authority. In this case, the champions at this level are ineffective in fulfilling their roles unless they are well connected with people at a higher management level. However, the situation becomes challenging when higher-level management people want to champion their projects. At the same time, an executive VP

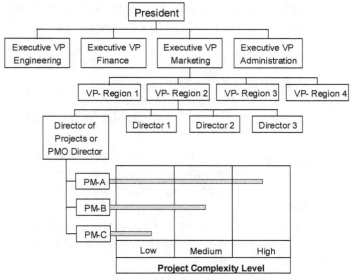

Figure 11.1. Relationship between project complexity level and positional power of a champion.

for a project C would be overkill. Politically sensible project managers try to find champions from an appropriate level in the management hierarchy so they can influence other management people and stakeholders as needed, to remove roadblocks project managers encounter.

11.1.3. Develop a strategy to deal with special issues by engaging outside consultants

> *"Good management consists in showing*
> *average people how to do the work*
> *of superior people."*
> —*John D. Rockefeller*

You, as a project manager, might be unsuccessful in implementing ideas, methodology, processes, and techniques in your organization because of politics, even if the ideas are sensible and good. However, if you get an outside consultant to suggest the same ideas, management listens to those ideas and even follows them.

Why is this true, and should we follow this technique? If you believe their ideas are good and can help the organization, then it is better to use outside consultants and facilitators to make things happen. Project managers should be primarily interested in results rather than worrying about their personal egos to prove themselves right. The best approach to deal with such political situations is to use outside consultants for their special knowledge and expertise.

Project managers can use outside consultants for two knowledge areas:

1. Subject matter experts — In this context, project managers should look for well-known consultants with a high reputation in knowledge areas related to issues the project managers want to resolve. Project managers can use them to support their ideas because top management might view outside consultants

as neutral and better subject matter experts.
They can push project managers' major ideas
in the organization. Politically, it is also a
win-win situation for the project manager for
the following reasons:

- If the consultants are well liked by
  management, and their advice is followed,
  project managers ultimately get the credit
  for finding a good consultant and getting
  things done.

- If consultants are not well received, and
  their advice does not lead to successful
  solutions, project managers might not be
  blamed as much as if they implemented
  similar ideas themselves.

  Politically, it is better for project
  managers because senior management
  made the final evaluation of the
  consultants and their advice, so they
  have to take responsibility for their
  actions and decisions. Furthermore, if
  senior management is unhappy with the
  consultants' performance, the project
  managers should back off pushing the
  ideas because there might be additional
  political backlash, and the timing might
  be inappropriate.

2. Special roles (lobbyists, arbitrators,
   agents) — In some projects and situations,
   project managers lack specific expertise
   and connections necessary to get things
   done. They should seek consultants to play
   special roles as agents, lobbyists, promoters,

and arbitrators, especially in dealing with international and global projects because of differences in culture, politics, economy, and regulations.

This help might be needed for projects involving large government departments such as the Department of National Defense, military, research, and regulatory authorities (pharmaceutical, medical, food, airlines, and so on) because of different ways of getting things done, which might involve submitting or filing applications for approval of certain products and services. Project managers should use well-known agents, lobbyists, and arbitrators to meet administrative requirements for submitting proposals, applications, appeals, funding requests, and so on.

These consultants are highly specialized experts in their areas and know the processes, shortcuts, and boundaries in detail. Therefore, working with them increases the confidence of senior management in you and your ability to find appropriate experts to achieve successful results. Working with consultants not only leads to better quality submissions, but also increases the likelihood of success and, hence, more positive overall feedback from senior management.

11.1.4. Find out What's in It for Them (WIIFT) to manage upward

> *The manager asks how and when;*
> *the leader asks what and why.*
>
> —Warren Bennis

Most project managers feel comfortable managing their project teams because they are established and informal and have a good working relationship with their team members. However, they lack confidence influencing senior managers and managing upward.

Project managers need senior management support to get resources, information, decisions, and approvals promptly to meet project deadlines. To meet this challenge, project managers must remember one of the three truths of life—most people do things to meet their interests—and then manage the senior managers.

Before lending their support, most people ask What's in It for Me (WIIFM)? While managing upward, project managers must keep this question in mind and address the project issues and problems from senior management's perspective. They must develop project solutions, benefits, and outcomes to focus on What's in It for Them (WIIFT). They should try to make their bosses look good by delivering successful projects for them, which increases senior managers' confidence in the project managers' abilities. Senior managers are happy to assign future high-profile opportunities to project managers with proper guidance and support.

11.1.5. Analyze political landscape and protect yourself from sharks

> *Learn the rules so you know how*
> *to break them properly.*
> —The Dalai Lama

Most project managers have excellent technical skills and people skills for managing their projects and their project teams. However, many project managers lack good political skills in dealing with their stakeholders in knowing what to say, how to say, and when to say. They are too straightforward, direct, and business-oriented. Politics are a necessary part of life in project environments. Therefore, project managers should analyze the political landscape to understand the following three political positions many people take in an organization:

1. Naives

2. Sharks

3. Politically Sensibles

Most project managers are challenged with understanding sharks, identifying them, and managing them effectively. As a basic political skill, they should understand that sharks are self-centered, and they look at politics as an opportunity to advance themselves, even if at the cost of others. They are excellent in hiding information, twisting information, and spinning information to their benefit. Sharks try to put obstacles and roadblocks in the way of project managers. As a first step, project managers should differentiate between the big sharks and baby sharks and then develop appropriate strategies to protect themselves from both types:

- Big sharks: These sharks are higher in the organization hierarchy and have much more positional power than project managers do. Direct confrontation is difficult. In such circumstances, project managers should use their champions, who are at a higher positional power than sharks are, to protect themselves from big sharks. The champions can use their formal and positional power to tell sharks to stay away from you.

- Baby sharks: In this context, baby sharks are stakeholders at the same or lower positional power as the project managers are. They can create small obstacles in planning and executing project activities that involve management of meetings, brainstorming and facilitating, conflict management and escalation issues, problem-solving and decision-making. In such circumstances, project managers should establish procedures and ground rules in consultation with stakeholders and then enforce the rules to discourage those who might interrupt them and create obstacles to getting things done.

11.1.6. Learn to be politically sensible! (Relate project goals to organizational and department goals.)

> *Believing everybody is dangerous; believing nobody is very dangerous.*
>
> —Abraham Lincoln

Politics can sometimes be viewed as a way to get your things done through others. Project managers must recognize that many projects are derailed, not because of technical reasons or poor project management processes or methodologies, but because of political issues. Therefore, project managers must learn to be politically sensible to manage politics at the project level and at the upper management level.

Project managers encounter big sharks and baby sharks along the way, but they should protect themselves by exposing sharks when they try to twist the information and create obstacles. Project managers must use the following two-step approach to be politically sensible:

1. Find organizational strategies.

    Project managers should learn the main organizational strategies and goals from their mentors and champions.

2. Align organizational strategies with project goals.

    Project managers should develop skills to show how their project goals align with organizational strategies and goals.

11 - Managing Politics at the Project Level

When asking for additional resources and support to expedite the project, project managers should relate the project problems and issues with organizational goals, department goals, and then project goals. They should articulate the project problems for their effects on organizational goals.

Most project managers have enormous responsibility, but little or no formal authority over their stakeholders. They face special challenges of managing upward and resolving various political issues to manage their projects effectively. Guidelines and tips to resolve such challenges include following the management hierarchy, gaining top management support by having good champions with appropriate positional power; developing a strategy to deal with special issues by engaging outside consultants; describing project benefits in the right language for executives to manage upward and discover What's in It for Them (WIIFT), analyzing the political landscape and protecting yourself from sharks, and learning to be politically sensible by aligning project goals to overall organizational goals.

Politics are inevitable in project environments. Project managers must develop skills to manage upward and deal with various political issues throughout the project lifecycle. They should first stop being naive, then identify and protect themselves from sharks, and finally become politically sensible by developing and using their power effectively and communicating with all stakeholders at all levels in the organization to gain their support and commitment continually.

11.2. Project Management and Team Leadership Issues

One determined person can make a significant difference, but a small group of determined people can change the course of history.
—Sonia Johnson

In project environments, project managers are expected to lead their project teams and develop good project management systems and processes to plan, execute, and monitor project activities efficiently. They must first develop informal powers that include referent, expert, information, network, and persuasion power, because they rarely have enough formal authority over their team members and other project stakeholders. Then, they must mostly use their power informally to influence their stakeholders and gain their support and buy-in.

They should look after their teams and create high-performance teams because project teams represent the most important power base of project managers. Practical tips to deal with politics related to project management and team leadership issues are shown in table 11.2.

Issues	Guidelines & Tips
Project Management & Team Leadership Issues	• Build a highly-committed and high-performance team. • Increase your overall power and use power effectively. • Help your team members increase their informal power. • Look after your team and yourself. • Use strategic instruments from the *PMBOK Guide*® as a guideline and escalate issues as necessary. • Mentor and coach your team to improve their people skills.

Table 11.2. Managing Politics at the Project Level (Project Management and Team Leadership Issues)

11.2.1. Build a highly committed and high-performance team

> *Most success depends on colleagues, on the team. People at the top have large egos, but you must never say "I"; it's always "we."*
>
> —Frank Lampl

Project managers' success depends on their team members' commitment and dedication. Therefore, project managers must pay special attention to their team members' needs. They must inspire their team members and create an environment to build a highly committed team, because the project team is their major power base and power source. The following are tips to build a highly committed team:

• Provide opportunities for training and development.

- Encourage cross training and sharing of information.

- Show confidence and trust in your team members.

- Reward appropriately to reinforce positively.

- Nurture creativity and innovation.

11.2.2. Increase your overall power and use power effectively

Use power to help people. For we are given power, not to advance our own purposes nor to make a great show in the world, nor a name. There is but one just use of power and it is to serve people.

—George Bush

Informal power—what you earn—is most helpful in getting things done. The following are advantages of informal power:

- No one can take it away from you, even if you change department or company.

- You can work to increase it immediately rather than waiting until you go up the management ladder or hierarchy.

- The more you share, the more you get.

- A solid foundation of informal power makes you a good candidate for promotion to formal positional power.

Therefore, project managers must work on increasing their informal powers (expert power, awareness of organizational policies and procedures and how the organization really works (information power), network power, leadership skills (referent power), and influencing skills to persuade your stakeholders (persuasion skills)). It not only increases your self-confidence, but also increases the confidence of senior management in you and your ability to get things done.

11.2.3. Help your team members increase their informal power

> *Unity is strength . . . when there is*
> *teamwork and collaboration,*
> *wonderful things can be achieved.*
> —Mattie Stepanek

Besides working on increasing their informal power, project managers should also help their team members increase their informal power. They can do so by proper coaching and mentoring. They should provide sufficient support for training and development by providing team members with opportunities to attend conferences, seminars, or courses, and cross training.

Proper coaching and mentoring increase team members' professional strength (technical and nontechnical) and increase their self-confidence in accepting new challenges and opportunities. Project managers must provide good leadership for their team members to help them grow, be motivated, and inspire them to enhance their performance. Team members have more respect and commitment for such project managers, and they do everything they can to collaborate more and minimize the negative impact of politics in meeting project goals and objectives.

11.2.4. Look after your team and yourself

> *Get the right people. Then no matter what all else you might do wrong after that, the people will save you. That's what management is all about.*
> —Tom DeMarco

Project managers must realize that their performance depends on their team members' performance. In this context, they must protect their team in front of other project managers and never talk negatively about their team members, especially about their weakness in technical and project management skills. Project managers should use the following practice: "When it comes to taking credit, give it to your team, and when it comes to taking blame, take upon yourself." This way, project managers gain more respect and cooperation from their teams.

In a way, project managers should provide coaching and training to their team members to enhance their professional strength. Project managers should also be careful when they face political situations and dilemmas and have to do things to support people and situations. Depending on their viewpoint and the extent of compromise they have to make in such situations, they must document the circumstances and reasons for their actions to protect themselves in the future, especially when they face difficult ethical dilemmas. In some cases, it is called Memo for Record (MFR). Project managers should have a political backup to get things done.

11.2.5. Use strategic instruments from PMBOK as a guideline, and escalate issues as necessary

Of all the things I've done, the most vital is coordinating the talents of those who work for us and pointing them towards a certain goal.
—Walt Disney

People management is more important than project management processes and methodology are. However, you have a better chance of making your case successfully if you back it up with a proper plan (schedule and resource analysis) and risk management strategies. Qualitative and anecdotal statements get attention, but project managers must use outputs from project management processes and analysis to support their case.

11.2.6. Mentor and coach your team to improve their people skills

The strength of the team is each individual member. The strength of each member is the team.
—Phil Jackson

The project team is the most important power base for project managers. Project success depends on the project team's performance. Project managers are responsible to optimize their project team's performance. Most project team members are competent in the technical skills of their knowledge area and in project management processes and tools. However, they lack people skills required to manage large and external projects.

They might not know the tricks of the trade, such as influencing without authority and political skills. Project managers should mentor and coach their project team members continually to improve their interpersonal skills. They should give them an opportunity to learn and use their people skills to deal with complicated situations and provide genuine support if they need it. Senior management should implement mentoring programs with the appropriate resources to achieve success.

Project managers must provide good team leadership and develop efficient project management processes, tools, and procedures to manage their projects effectively. Project managers should follow practical guidelines to manage project politics related to team leadership and project management issues: increasing their informal power base and using their powers effectively, helping team members increase their informal power, developing skills to build effective and high-performance teams, looking after their team by giving them proper rewards and recognition, developing efficient processes and techniques to execute and monitor project activities and escalating issues when necessary, and mentoring team members to improve their project management skills and people skills.

Effective team leadership and project management systems are crucial to project success. Project managers must focus on increasing their influencing skills and team-building skills to increase the performance of all team members. They should protect their teams and themselves from negative organizational politics. They should continually coach their team members about organizational politics so they can anticipate political issues to minimize the risk of serious project problems. They should use the PMBOK as a guideline to develop efficient project management tools and techniques, processes, and templates to monitor the project, and produce effective status reports and warning signals to keep clients and senior management well informed, and develop solutions to manage project risks in time.

11.3. Stakeholder Management Issues

> *A true architect is not an artist but an optimistic realist. They take a diverse number of stakeholders, extract needs, concerns, and dreams, and then create a beautiful yet tangible solution that is loved by the users and the community at large. We create vessels in which life happens.*
> —Cameron Sinclair

Most projects have stakeholders with different interests, concerns, values, and knowledge. A stakeholder can be an individual, group, or an organization that might affect, be affected by, or perceive itself affected by a project's decision, activity, or outcome. As one primary responsibility, project

managers are expected to manage project stakeholders and keep them happy.

A significant component of managing politics at
the project level involves effective management of
various stakeholders. There can be different definitions of project success by different stakeholders.
The project might be considered unsuccessful, even
if the project deliverables are met, and the project is
completed on schedule and within budget. Whenever
project managers ask for support from stakeholders,
they must ask why they should they help and What's
in It for Them (WIIFT)?

Project managers should differentiate between
the motives and expectations of stakeholders. They
should communicate with their stakeholders by
keeping these questions in mind and then explaining
how the project benefits meet their interests and
concerns. Guidelines and tips to deal with politics related to stakeholder management issues are shown
in table 11.3.

Issue	Guidelines & Tips
Stakeholder Management Issues	• Identify key stakeholders and develop strategies to manage them. • Analyze five categories of stakeholders based on trust and agreement model and then manage them appropriately. • Remember three simple truths of life and manage stakeholders accordingly. • Ask management to clarify goals, roles, and responsibilities to avoid confusion. • Resolve communication and conflict problems effectively at all levels. • Practice effective filtering of information.

Table 11.3. Managing Politics at the Project Level
(Stakeholder Management Issues)

11.3.1. Identify key stakeholders and develop strategies to manage them

> *Find the appropriate balance of competing claims by various groups of stakeholders. All claims deserve consideration but some claims are more important than others.*
>
> —Warren Bennis

Most projects have stakeholders with different interests, expectations, and personalities. Stakeholder management is important to managing politics at the project level, and it is critical to overall project success.

The project manager is the most important stakeholder responsible for overall project success. Stakeholder management includes processes to identify stakeholders, analyze stakeholder expectations and their impact on the project, and develop strategies to manage stakeholders effectively to gain their support (PMBOK Guide 2013, 391–392).[2] As a first step to manage stakeholders, project managers should identify different categories of stakeholders. Verma (1995, 46–70) described major categories of stakeholders:[3]

- Client or project sponsor or champion.

- Project team.

- Internal stakeholders: These stakeholders are directly related to the project and include top management, functional managers, staff and service personnel, and other project managers.

- External stakeholders: These stakeholders are indirectly involved with the project, and they can indirectly influence project outcomes. External stakeholders must be given special attention because of their concerns, which include regulations—environmental and legal, public and press, social and cultural, economic and financial.

- Contractors, subcontractors, and major vendors: These stakeholders can be identified as external or internal stakeholders.

- Competitors, associations, and other vested interest groups.

Project managers should recognize the roles and abilities of all stakeholders to influence project outcomes.

Tres Roeder (2013, 31–37) provided excellent tools and tips for analyzing and managing project stakeholders in his book *Managing Project Stakeholders.*[4] Project managers must develop appropriate strategies to manage their stakeholders throughout the project life cycle (Roeder 2013, 31–37; Muhammad 2013):[5]

- **Analyze stakeholder expectations —** From a political viewpoint, project managers must differentiate between motives and expectations. Expectations are gathered and ranked by their importance during the project planning phase. Sometimes, expectations are unclear and keep changing throughout the project life cycle, which especially challenges project managers when they try to meet or exceed stakeholders' expectations.

Motives are internal, and they are rarely discussed. Unlike expectations, motives are the real drivers behind what stakeholders say and how they think, act, and behave. The consistencies between words and actions determine the actual motives (Muhammad 2013).[6] For example, some stakeholders could be verbally supportive in a face-to-face conversation, but they do not do their best to help the project. The same stakeholder in management meetings might be unsupportive in both words and actions. Project managers should observe stakeholder behavior in different situations to understand their motives and then develop proper strategies to influence them to gain their genuine support.

- **Prioritize stakeholders** — Not all stakeholders are equal. Project managers should prioritize their stakeholders by their importance, power, and influence, and deal with them differently. Four attributes by which to prioritize stakeholders are the following (Roeder 2013, 31–37):[7]

  1. **Power:** Positional power of stakeholders or authority levels (high or low)

  2. **Impact:** Ability to effect changes based on their knowledge (high or low)

  3. **Interest:** Level of concern regarding project outcomes (high or low)

  4. **Influence:** Level of active involvement (high or low)

Based on relationships among the four attributes, project managers can determine the effort needed to manage various stakeholders, knowing which stakeholders must be managed closely, who should be kept informed, and who should be monitored.

- **Build alliances with key stakeholders —** Keeping all stakeholders happy is difficult. However, project managers must identify key stakeholders with high power, influence, knowledge, and respect in the organization. Project champions should certainly be considered key stakeholders, and they must be involved in the early parts of project planning to remove project obstacles. Project managers should continually analyze the degree and the depth of their relationships with different stakeholders and departments to gain support throughout the project life cycle. They should develop their network power by nurturing and feeding the network, that is, by helping others and maintaining quality relationships.

- **Gain support from stakeholders —** Project managers have enormous responsibility, but no direct authority over their team members and project stakeholders. Therefore, they must earn, rather than demand, support from stakeholders. Earning buy-in and support is a process, not an event. Project managers should use this process continually to enlarge their circle of support by including stakeholders, observing stakeholders, and responding to them quickly with sensitivity to their feelings and beliefs throughout this process. They should use a collaborative approach and help stakeholders find the best solution for themselves and for the project. This way, it seems as though what they are doing is in their best interest (Roeder 2013, 31–37).[8]

- **Act like a leader** — Project leadership and project management are both essential in delivering successful projects. Project managers should play the role of a leader, as needed, and learn the art of balancing both to achieve success. They should lead stakeholders to manage them effectively. They should use the concept of sixth sense for project management developed by Tres Roeder (2011), which is a portfolio of the following six disciplines required to be an effective leader:[9]

1. Awareness

2. Whole-body decisions

3. Clear communications

4. Adaptability

5. Diplomacy

6. Persistence

Based on the six disciplines of a sixth sense of project management, Tres Roeder (2011) suggested a three-step approach for effective leadership:[10]

1. Be aware of yourself, others, and situations.

2. Adapt according to people and situations to lead effectively.

3. Act, which includes the disciplines of communicating clearly with diplomacy and persistence and using your brain, heart, and gut feelings to make good decisions.

* **Choose your fights** — Project stakeholders often disagree. Project managers might face challenges in dealing with stakeholders who have opposing positions and viewpoints about project management systems and approaches.

  Some stakeholders take project management issues personally and do everything they can to create obstacles for project managers. In such cases, project managers should analyze the situation and understand the motives behind the way stakeholders think, act, and behave. They should pick the battles critical to project success that can be won. They should establish boundaries for acceptable behavior and support from stakeholders and not become involved in fights as long as the stakeholders stay within those boundaries. Project managers should escalate the issues only when stakeholders cross the boundaries of acceptable behavior and support.

11.3.2. Analyze five categories of stakeholders based on trust and agreement model and then manage them appropriately

Remember, teamwork begins by building trust. And the only way to do that is to overcome our need for invulnerability.

—Patrick Lencioni

Project managers must develop strategies to build support for their visions and their projects. Peter Block (1987, 137–160) developed a model based on the following two dimensions that determine whether our stakeholders support us as our allies or do not support us and become our adversaries.[11]

1. Trust: implies confidence, faith, and reliance

2. Agreement: implies harmony, concurrence, and mutual understanding

Project stakeholders can be divided in five categories based on the two dimensions, as shown in figure 11.2 (Block 1987, 137–160).[12] Project managers should aim to convert as many stakeholders as possible into allies.

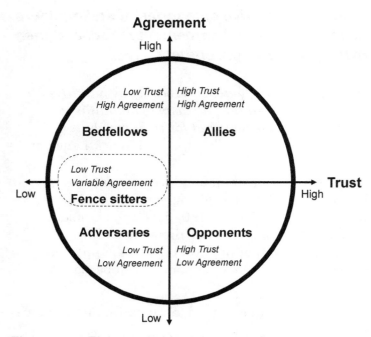

Figure 11.2. Five categories of stakeholders (based on trust and agreement).

The following describe the five categories of stakeholders with strategies to manage them (Verma 2012–2015; Block 1987, 137–160):[13]

1. Allies (High Trust/High Agreement)— Mentors and Champions

   Allies share our vision and want us to succeed in our way. They can be trusted to discuss our strengths and weaknesses because they want to help us progress in the organization. In a project life, mentors and champions are our best allies. The following are tips to deal with allies:

- Involve allies early in your project, as much as possible.

- Help allies in meeting their goals.

- Thank allies for their help and cultivate relationships long term.

In a project environment, our mentors and champions should be our best allies because we depend on them to remove our project obstacles. We should confirm the quality of relationships and maintain it.

2. Opponents (High Trust/Low Agreement)— Positive Devil's Advocates

Opponents can be trusted, but they might disagree with us about our mission and plans. They introduce new challenges regarding our plans and techniques, intending to improve them and make them more realistic. Project managers should not ignore their viewpoints without thorough analysis and evaluation because you might achieve creative and innovative solutions by following their suggestions. The following are tips to deal with opponents:

- Be positive and grateful for their challenges rather than rejecting them.

- Confirm the quality of relationship and that it is based on trust.

- State your vision regarding the project objectives.

- Stage their position and viewpoints neutrally.

- Use a problem-solving approach on a one-to-one basis to reach agreements.

3. Bedfellows (Low Trust/High Agreement)— Yes People

   Bedfellows are stakeholders who agree with you in face-to-face meetings about your plans and how to proceed, but they cannot be trusted for their support. You must acknowledge the caution that exists from experience. The following are tips to deal with bedfellows:

   - Be assertive with your goals and plans.

   - Be clear about what you want from them in working together and ask bedfellows to be clear about what they want from you.

   - Use a third party as evidence, and document the agreement. Volunteer to write the agreement with the implementation plan to emphasize your position, because the person with the pen has the power.

   Bedfellows can be difficult to manage effectively because they yes-yes to your face, but when they turn around, they behave differently. Project managers should let bedfellows express their positions clearly to understand their expectations and learn how to gain their goal cooperation.

4. Fence Sitters (Low Trust/Variable
    Agreement)

    The agreements reached with fence sitters
    are variable and unknown, and they cannot
    be trusted for their commitment to carry
    through the agreements. They cannot take
    a stand for us or against us. They are over
    noncommittal and keep bringing risks and
    uncertainties into the picture. However, they
    might be great at presenting themselves as
    practical, mature, and cautious. The following
    are tips to deal with fence sitters:

    *   State your position clearly and ask fence
        sitters where they stand.

    *   Use third party as evidence, and docu-
        ment the agreements when reached.

    *   Apply gentle pressure by reminding them
        what you have done for them lately.

    *   Express disappointment for their lack of
        genuine support.

    *   Ask them what it would take to gain their
        support (leave doors open).

    Politicians and bureaucrats can be
    considered typical fence sitters because they
    feel insecure in making decisions because
    of lack of in-depth knowledge and self-
    confidence. They keep asking for more and
    more data, studies, and analysis to justify
    their procrastination.

5. Adversaries (Low Trust/Low Agreement)

Adversaries are difficult to manage because of low agreement and low trust. People become adversaries when your repeated efforts in negotiating agreements and trust fail. In a way, they can be considered high-maintenance people because it requires extensive effort to reach agreements and gain their support. The following are tips to deal with adversaries:

- Do not label people as adversaries too quickly. Give them the benefit of the doubt, and try to see things from their perspective.

- State your vision, and explain how it aligns with organizational goals.

- Acknowledge their position neutrally without raising defensiveness.

- Identify your contribution to the problem in the past and apologize, if necessary, with a request to let go and proceed with a clean slate.

- End meetings with your goals and plans and no demands from them.

Managing stakeholders by enhancing trust in agreement is an excellent and practical approach. Project managers must identify project stakeholders and differentiate between the motives and expectations.

11.3.3. Remember three truths of life, and manage stakeholders accordingly

> *You can have everything you want in life*
> *if you just help enough people get*
> *what they want in life.*
> —Anonymous

Project stakeholders are normal people who like to be cared about and appreciated. They are motivated or demotivated depending on how the project manager and team members treat them. Project managers should treat their stakeholders the way they want to be treated themselves. They should remember the following three truths of life, and then manage their stakeholders accordingly (Verma 2012–2015).[14]

1. People make or break things — People make things happen or prevent things from happening. Project managers must recognize that a project's success depends on how project stakeholders think, behave, and act. Project managers should pay special attention to their stakeholders' concerns and interests and address them promptly to gain their cooperation.

2. People do mostly what is in their best interest — Most people, when requested to help others on the project, ask What's in It for Me (WIIFM)?

   Project managers must recognize this question's importance and address it before expecting support from their project stakeholders, which helps project managers see things from the stakeholders' perspective.

Project managers should use a strategy of describing why stakeholders should help the project and how it would satisfy their interests. They should describe the potential benefits of outcomes resulting from stakeholders' actions for What's in It for Them (WIIFT) to gain genuine support from their stakeholders.

3. People support what they create — Most people like to be involved in creating things they support. Project managers should remember this and involve their project stakeholders in project planning and execution. They should make their stakeholders part of the creation process for preparing project plans, budgets, risk analysis, and project reports and making important decisions. This involvement convinces stakeholders to own project issues and problems because they were intimately involved in most project activities and initiatives.

The truths of life mentioned here are simple but powerful in gaining cooperation from stakeholders. Sometimes, stakeholders behave and act differently, compared with what they say in a face-to-face meeting, which creates confusion and misunderstanding between the stakeholder and project manager and leads to negative politics. Effective management of stakeholders' politics involves remembering the three truths and then dealing with them accordingly.

11.3.4. Ask management to clarify goals, roles, and responsibilities to avoid confusion

The father who does not teach his son
his duties is equally guilty
with the son who neglects them.
                                    —Confucius

Clear definition of project goals, project plan, and execution strategy are essential to project success. Lack of clarity in these areas leads to confusion, misunderstandings, and breakdowns in communication. Project stakeholders might act on their assumptions, causing project problems. In such cases, stakeholders might blame others, leading to negative politics.

Project managers must start the project on the right foot by establishing clear and realistic project goals and defining roles and responsibilities of all project stakeholders. They should do so by involving all necessary project stakeholders to gain their buy-in and commitment to work together in meeting project goals and objectives.

Project managers must assign roles and responsibilities to the right stakeholders to ensure project success. According to Verma (1995, 99–102, 119–120), responsibility without authority leads to frustration and poor performance. When responsibility is assigned to certain stakeholders, proper authority should also be assigned to them to go with it. Reliability and accountability are interrelated in that reliable people feel accountable on their own. Reliability encompasses record of accomplishment and quality of work. Therefore, project managers should assign responsibilities to stakeholders and expect high reliability from them. At the same time, they should assign them the proper authority to go with these responsibilities and hold them accountable for the outcomes.[15]

In particular, project managers should use the following guidelines related to this area to avoid potential problems:

- Align project goals with organizational goals to gain support from all stakeholders.

- Use the SMART goal-setting approach to establish project goals; that is, project goals should be (Verma 1997, 122–123, 216)[16]

  - Specific (to provide focus and avoid confusion)

  - Measurable (to monitor progress)

  - Achievable (be realistic but challenging)

  - Results-oriented and rewardable (provides a sense of achievement and motivation)

  - Time-based (should have clearly established milestones and schedule)

    Project managers should establish clear goals to provide purpose and direction for project team and project stakeholders. This effort avoids typical problems in establishing goals related to issues, which include ambiguity, conflict, overload, and complexity.

- Assign roles to stakeholders according to their skills and capacity, clarify roles to ensure nothing falls through the cracks, and provide necessary support to ensure success.

Assign responsibilities to stakeholders and give them proper authority to carry out their responsibilities successfully. To ensure success, project managers should assign responsibilities with proper authority and expect reliability and accountability from stakeholders.

11.3.5. Resolve communication and conflict problems effectively at all levels

> *Any fool can criticize, complain, and condemn—and most fools do. But it takes character and self-control to be understanding and forgiving.*
> —Dale Carnegie

Often, project stakeholders have different interests, expectations, and backgrounds. Communication problems and conflicts among stakeholders are common at all levels in project environments. These problems can lead to misunderstanding, frustration, stress, and a lack of teamwork, leading to negative politics. Project managers must look for verbal and nonverbal clues to identify communication problems and potential conflicts and resolve them effectively to avoid major project problems.

The following are barriers and bottlenecks to successful communications project managers should identify (Verma 1996, 24–25, 142–143):[17]

- Amount of information. (Keep it simple and avoid overload or underload.)

- Lack of knowledge. (Stakeholders should be trained and coached.)

- Cultural differences. (Encourage knowledge about different cultures and ethnic backgrounds.)

- Number of links. (Keep to a minimum.)

- Perceptions and personalities. (Encourage to see others' perspectives.)

- Message competition. (Get full attention when communicating.)

- Project jargon and terminology. (Use simple words and avoid jargon.)

- Organizational climate. (Avoid status or ego issues. Encourage stakeholders to talk *with* people rather than down to people.)

Conflicts happen when stakeholders have incompatible goals, interests, thoughts, or emotions, and when working together, they act and make decisions to meet their objectives rather than overall organizational objectives. Effective conflict management focuses on minimizing the negative impacts of conflicts. Project managers must identify conditions and reasons leading to conflicts, and then understand various conflict resolution techniques to choose the best one, depending on time constraints, personalities involved, and overall impact on project performance. The following are five common interpersonal conflict resolution techniques:

- Withdrawing/Avoiding
- Smoothing/Accommodating
- Forcing/Imposing
- Compromising/Bargaining
- Confronting/Problem-Solving/Collaborating

Project managers should use the following three-step approach to manage conflicts in a project environment (Verma 1996, 24–25, 142–143):[18]

1. Prepare for the conflict (by acknowledging and planning).

2. Face the conflict (by listening and responding to both emotions and facts, finding the real issues and analyzing the situations).

3. Resolve the conflict (by looking for win-win solutions, cutting losses when necessary, and developing strategies to prevent and handle conflicts).

11.3.6. Practice effective filtering of information

*The most important thing in communication
is hearing what isn't said.*
—Peter Drucker

Often, project managers are responsible for gathering, analyzing, editing, and disseminating information to project stakeholders. Project stakeholders want the information in some detail, style, and format. However, project managers might decide not to disseminate all information to all stakeholders.

They might filter information before disseminating it to some stakeholders because the information might be classified, sensitive, and confidential.

Sometimes, they summarize and disseminate only the highlights of certain project reports to avoid information overload. In this context, project managers must evaluate the negative and the positive impacts of filtering information given to project stakeholders, observe their responses, and then filter the information accordingly. Project managers should be open and straightforward when communicating with stakeholders and let them know they need more time to analyze and evaluate the information before disseminating it to them and making it in the right format to help them do their job more effectively. This prevents unnecessary complaints from stakeholders about not getting the full information.

Stakeholder satisfaction is the key to project success. Stakeholder management includes the processes to identify stakeholders, analyze their expectations and their impact on the project, and develop strategies to gain their support and buy-in continually. Successful stakeholder management is an important responsibility of project managers.

Project managers must understand different categories of stakeholders (internal and external), analyze their expectations, and prioritize them based on various factors: positional power, impact on the project (ability to effect changes), interest (level of concern), influence (level of active involvement), and knowledge (about the project and related industry). Project managers should build alliances with key stakeholders. They should identify and analyze the five categories of stakeholders—allies, opponents, bedfellows, fence sitters, and adversaries— based on the dimensions of trust and agreement. They should try to convert as many stakeholders as possible into allies to maximize support for their vision and project management approach.

Other guidelines include remembering the three truths of life and then managing the stakeholders accordingly; asking senior management to clarify goals, roles, and responsibilities to avoid any confusion among stakeholders; resolving communications and conflict problems at all levels effectively; and practicing effective filtering of information.

Chapter 11 Summary

Project managers must understand different categories of stakeholders (internal and external), analyze their expectations, and prioritize stakeholders based on various factors:

- Positional power
- Impact on the project (ability to effect changes)
- Interest (level of concern)
- Influence (level of active involvement)
- Knowledge (about the project and related industry)

Most project managers have enormous responsibility, but little or no formal authority over their stakeholders. They face special challenges in managing upward and resolving various political issues to manage their projects effectively. The following are guidelines and tips to resolve such challenges:

- Follow management hierarchy.
- Gain top management support by having good champions.
- Develop a strategy to deal with special issues by engaging outside consultants.
- Find out What's in It for Them (WIIFT) to manage upward.
- Analyze political landscape and protect yourself from Sharks.
- Learn to be politically sensible! (Relate project goals to organizational and department goals.)

Politics are inevitable in project environments. Project managers must develop skills to manage upward and deal with various political issues throughout the project lifecycle. They should first stop being naive, then identify and protect themselves from sharks, and finally become politically sensible by developing and using their power effectively and communicating with all stakeholders at all organizational levels to gain their support and commitment continually.

Project managers must provide good team leadership and develop efficient project management processes, tools, and procedures to manage their projects effectively. Project managers should follow practical guidelines to manage project politics related to team leadership and project management issues:

- Build a highly committed and high-performance team.

- Increase your overall power and use power effectively.

- Help your team members increase their informal power.

- Look after your team and yourself.

- Use strategic instruments from the *PMBOK Guide*® as a guideline, and escalate issues as necessary.

- Mentor and coach your team to improve their people skills.

Effective team leadership and project management systems are crucial to project success. Project managers must focus on increasing their influencing skills and team-building skills to increase the performance of all team members. They should protect their teams and themselves from negative organizational politics. They should continually coach their team members about organizational politics so they can anticipate political issues to minimize the risk of serious project problems. They should use the *PMBOK Guide*® as a guideline to develop efficient project management tools and techniques, processes, and templates to monitor the project, and produce effective status reports and warning signals to keep clients and senior management well informed, and develop solutions to manage project risks in time.

To manage stakeholders and deal with related issues, project managers should use the following guidelines:

- Identify key stakeholders and develop strategies to manage them.

- Analyze five categories of stakeholders based on trust and agreement model and then manage them appropriately.

- Remember three truths of life, and manage stakeholders accordingly.

- Ask management to clarify goals, roles, and responsibilities to avoid confusion.

- Resolve communication and conflict problems effectively at all levels.

- Practice effective filtering of information.

Part IV Summary

Project managers must understand the three truths of life outlined in this part. These truths include the how important people are, what turns them ON to do project activities in certain ways, and what is necessary to gain support from various project stakeholders.

People make or break things is the first truth. In other words, people make things happen and prevent things from happening, that is, people do projects, and therefore, people skills are critical to deliver successful projects. People behave differently under different circumstances and at an individual level, team level, or management level.

People do mostly what is in their best interest is the second truth. Before people do things to help others, they look for an answer to the question What's in It for Me (WIIFM)? Therefore, to manage stakeholders and especially to manage upward, project managers should sell their projects by explaining the project benefits in terms of the stakeholders' interests, that is, What's in It for Them (WIIFT).

People support what they create is the third truth. Therefore, the key to gain support from project stakeholders for any initiatives is to make them integral to the creation process so they take ownership and feel committed to successful outcomes. In this context, people's involvement and participation should be genuine, and they should feel free to express their opinions and concerns. To achieve success in implementing initiatives, project managers should ensure that agreements reached meet three conditions: 1) the agreements must be free of fear, threat, or coercion, 2) agreements must be fair and reasonable, and 3) there must be a meeting of the minds.

Politics are inevitable in organizations, especially in project environments. Project managers cannot afford to ignore or avoid politics because doing so can derail a project despite good planning and project management processes and tools. Project managers must learn the art of managing politics at both the senior management level and at the project level. They must become politically sensible to gain the support of management and project stakeholders throughout the project lifecycle.

Politics at the senior management level are very important, and they must be managed at the front end.

Part IV Summary

These politics are associated with three issues:

1. organizational issues
2. leadership issues
3. project management issues.

The management of politics at the project level are challenging because of the behavioral dimensions of project stakeholders. Politics at the project level are also associated with three types of issues: 1) political issues and the challenges of managing upward, 2) project management and team leadership issues, and 3) stakeholder management issues.

Appendix: Case Studies & Exercises

Case Study, Part I: Dynamics of a Troubled Project

Exercise for Part II: Analyzing Political Landscape to Manage Stakeholders

Case Study, Part III: Political Issues with Filtering Information

Case Study, Parts III & IV: Swimming with Sharks

Case Studies & Exercises

Case Study, Part I: **Dynamics of a Troubled Project**

An organization—ABC Limited—has bid on large government projects for the last two to three years. This year, ABC Limited submitted a bid on a project for design and maintenance of a database for motor vehicles used by the state government's Vehicle Insurance Corporation. The project was $5 million, and ABC Limited won the contract.

The organization was happy, and they had a celebration party and enjoyed an "enthusiasm" phase that lasted two weeks. After two weeks of celebrations, the team met to go through the project requirements, scope of work, the deliverables, and project schedule in detail. The project manager and the team members rolled up their sleeves and prepared a detailed project plan.

After presenting the plan to the steering committee, everyone was disillusioned and concerned
because it required considerable effort to design,
develop, and complete the deliverables on time. The
estimated budget seemed low, the scope looked much
larger than the team had thought, and not enough
people with the proper skill mix were available to
complete the project on time and on budget. The
team was confused, disillusioned, and depressed for
one to two weeks (phase 2—the "disillusionment"
phase).

They panicked, assigned the tasks to the people
available, and worked hard to fast track the project
in phase 3, the "panic" phase. Everyone worked
harder, not smarter, and the consequences became
obvious. In this panic phase, many mistakes were
made, quality was low, and important items were
overlooked. Many irrational and wrong decisions
were made, leading to extra work, rework, and
inefficiencies. Everyone working on the project felt
stressed and did not want to take full accountability
for project problems, leading to the fourth phase of a
troubled project—"searching for the guilty"—which
is not only nonproductive, but also destructive to
teamwork.

The project manager had read in many books
that as soon as this happens, people become even
more defensive and play political games. As a
result, the innocents often are punished, and the
people who never participated get the praise and
honor, which is exactly what happened in this case.
Innocent team members who worked hard to get the
project on track were punished (the fifth phase), and

the few who contributed little to advance the project received praise and honor (the sixth phase).

Questions:

1. How would you identify and describe this troubled project's six phases? What are the possible causes of each phase?

2. How would stakeholders feel emotionally during each phase and why? What drives each of this project's six phases when handled emotionally?

3. What should project managers do to avoid the negative impact of each phase of this troubled project and convert the emotional views into proactive views? What drives each of this project's six phases when handled proactively?

Exercise for Part II: Analyzing Political Landscape to Manage Stakeholders

To understand the political dynamics among stakeholders, you must analyze the organizational structure from a political viewpoint. Analyzing the political structure helps recognize informal relationships critical in project execution and implementing required changes. As a first step, project managers should identify main characteristics and political features of various stakeholders' roles and identify the names of stakeholders associated with these characteristics with an associated symbol on the organization chart. However, they should be cautious to keep it confidential rather than sharing with everyone. The following are main characteristics of stakeholders to analyze the political context around the project:

Characteristics for political analysis of organization structure:

1. True allies or project champions (very important; symbol: SSS)

   These include stakeholders sold on the project who would advocate for support from management and other stakeholders. These stakeholders are very important because they help remove roadblocks. They might challenge you to make your plans more effective.

2. Strong supporters (important; symbol: SS)

   These stakeholders are very strong. They prevail, win, and get their way with senior management and other managers. They prevail through charm, charisma, and influence or through bargaining or force.

3. Knowledgeable (about the project and the
   organization; symbol: K)

   These stakeholders are subject matter
   experts and very knowledgeable about
   the project and organization. They know
   organizational policies and procedures.

4. Winners or losers (symbol: WL)

   These stakeholders are likely to gain or lose
   the most from success or failure of the project.
   It refers to whose department is likely to
   have a major impact by the project.

5. Authority (symbol: A)

   These include stakeholders with formal
   authority over team members.

6. Influencers (very useful; symbol: I)

   These include stakeholders who are respected
   and who can influence the decision-makers.
   They know power brokers and different
   personalities of stakeholders. They are
   typically involved in major decisions.

7. Approachable and Accessible (symbol: O)

   These approachable stakeholders are
   generally accessible to you whenever you
   need them to discuss project issues and
   strategies. They are willing to help meet
   project objectives.

8. Decision-makers (very important; symbol: D)

   These stakeholders can make decisions about
   committing financial and human resources
   to the project. They know when and how

to deviate from typical and conventional company policies to help move the project forward.

9. Danger (symbol: XXX)

   These stakeholders are likely to damage the project and the project manager. Project managers should be extra careful about these stakeholders because they might act as sharks and play negative politics behind the scenes to sabotage the project.

10. Fence sitters (symbol: FS)

    These stakeholders are politicians and bureaucrats who do not take a stand for or against you. They are over cautious and noncommittal and keep bringing risk and uncertainty.

11. Adversaries (symbol: X)

    These stakeholders are difficult to manage because they are not open-minded and often disagree on every point. Your personalities do not mesh. They typically behave as adversaries and do not offer genuine help to meet project objectives.

Questions:

1. Review your company's organizational structure, identify main stakeholders with one or more of the above characteristics, and mark them with the above-explained symbols.

2. Review the roles of these stakeholders and describe how you would deal with each to manage politics effectively?

3. Analyze your political landscape to identify your allies and adversaries. How would you convert adversaries to allies?

Case Study, Part III: Political Issues with Filtering Information

Creative Solutions Limited (CSL) was well known for managing their projects effectively. John Dickson had worked for more than ten years as a project manager, and he had excellent people skills. He was well connected in CSL, and he kept his eyes and ears open about political issues because he had seen that many projects could be derailed because of political reasons, despite good planning and project management processes. He was working on a high-profile project to analyze and increase market share in Western Europe and Australia. This project had a tight schedule of fifteen months and a budget of $1.2 million.

John had several difficulties in getting the proper funding and people with the right skill mix to complete this project on schedule and on budget. James Thomas, the senior vice president of IT and sales support was interested in this project and acted as a champion for it. He regarded John Dickson highly, and he was happy John Dickson was assigned to manage this project.

John made several presentations to James and then to the Steering Committee to fight for higher priority and required resources. The Steering Committee met every two months to evaluate all projects and their priorities. Sometimes, it meant fast tracking or delaying projects and reallocating resources, as needed, to meet organizational objectives. John was happy that James got support from management, as needed, to help him manage the project effectively.

John had a committed project team with members from various functional departments working on a matrix basis. John addressed the major problems and concerns of team members promptly, and he created good synergy and collaboration. He met with the project team regularly to monitor progress and identify risks that might delay the project schedule and increase costs. He met with James at least every month, or more frequently if needed, to give him the project's status and point out any risks and flags that might affect the project. He noticed that James showed keen interest in the project and asked if further help or support was needed to ensure the project objectives were met effectively within budget and schedule constraints.

The project was going fine, and John met all major milestones up to the eleventh month. One day, John was preparing his presentation for the steering committee to present a comprehensive review and status of this project. He had collected all required information from all work package managers to report the status on the schedule and budget with the forecast to complete the project.

Just half an hour before the meeting, he received a call from Roger Fisher, production supervisor of a major vendor from whom John had ordered a server to test the software system the project team developed. Roger told John they had production problems, and the server would be three to four weeks late. John was shocked and disturbed by this information because he had just talked a week ago with the senior manager of production who told him everything was going well, and their server would be delivered on time. John could not spend much time evaluating this information from Roger, as he still had a few things to finish for his presentation at the Steering Committee meeting.

He finished his presentation, but he kept wondering how to handle the information about the server in the Steering Committee. He knew the server was crucial to test the software system, but he did not have enough time to discuss this information with his scheduling team and evaluate its impact on the overall schedule.

John thought about the political dynamics in the Steering Committee. Other project managers had told him to watch out for Ken Gaty and Bob Shaw who could act as sharks and play negative politics to promote their interests. He knew they did not like him and did not support his project well. Ken wanted someone other than John to manage this project. John was confused about whether to tell this information to the Steering Committee and, if so, how. He worried about the situation and wondered how to handle it and related politics effectively.

Questions:

1. Should John tell the Steering Committee members about the possible delay in server delivery?

2. If so, how should he share this information with the Steering Committee?

3. If not, then how should he handle this situation with the Steering Committee?

4. What are the related political issues about telling or not telling this information?

5. How could John manage the political issues?

Case Study, Parts III & IV: Swimming with Sharks

This case study illustrates dynamics and behaviors of Naives, Sharks, and Politically Sensibles.

Island Hydro (IH) is a major utilities company on Vancouver Island in British Columbia, Canada. IH is interested in outsourcing the development of a software system to balance power distribution among large industrial customers. This project, Software for Balancing Power Distribution (SBPD), with a budget of $25 million and a five-year schedule, was high profile for the provincial government because it would save overall power consumption.

IH prepared the request for proposal (RFP) and sent it to ten software consulting/development organizations. Five consulting firms, including Island System Services Ltd. (ISSL) submitted their bid, and ISSL was selected. ISSL was excited to win the project and wanted to use it to establish their reputation in software development for large hydro and utilities companies.

The president and CEO of ISSL appointed Nikhil, an executive vice president, as a sponsor/leader for the project. Nikhil has worked with ISSL for many years. He is well respected by senior management and the project management group at ISSL.

A steering committee consisting of ISSL board members (Jaya, Reyva, Rahi, and Veeyan), the president and CEO, and some senior staff members of ISSL was formed to monitor the project carefully because of its high profile and importance for their future reputation. Nikhil was given one month to prepare an overall project plan with proper risk analysis. Nikhil started on this right away, prepared a high-level plan with required resources, and appointed Rohnik as project manager for the project (SBPD). Nikhil emphasized its importance and asked him to prepare a detailed project plan with schedule, budget, risk analysis, and communication plan and authorized him to assemble the project team to deliver the project successfully.

Nikhil has worked with ISSL for twenty-five years and mentored many project managers, including Rohnik. He is well respected in the company for his wide industrial experience, and he is renowned in software development for hydro. Rohnik worked on many projects with Nikhil and delivered successful outcomes every time. Nikhil is impressed with Rohnik for his hard work, perseverance, people skills, high integrity, ethics, and straight forwardness.

Rohnik is technically competent with good project management and people skills, and he has developed expertise for software systems for utilities. At ISSL, he has managed three projects successfully for different sponsors, including one for Nikhil. Rohnik highly respects Nikhil for his knowledge and leadership skills.

Rohnik formed a project team and presented his project plan with detailed budget, schedule, and risk analysis to the Steering Committee. Glen challenged Rohnik on many points and Rohnik answered his questions and concerns. The Steering Committee approved Rohnik's project plan with presented budget, resource analysis, and schedule. However, Rohnik has become increasingly aware that many managers in ISSL are self-centered political sharks who are interested in meeting their objectives even at the cost of others in the organization.

Glen has twenty years of software industry experience, and he was hired four years ago by ISSL at the senior director level as a head of the Software Quality Assurance Department. He is clever, and he wants to go up the management ladder in the next two to three years. He is also a member of the Steering Committee responsible to oversee the SBPD project. He wanted Alan, his favorite, to become the project manager for the SBPD project to promote himself and his department. Glen knows some senior managers in IH and plays golf with them occasionally. Alan and Glen worked at Universal Systems together for four years before, and now, Glen is Alan's immediate boss in ISSL.

Alan is ambitious and wants to follow a project management career track. Although he has strong technical knowledge and more experience than Rohnik, he lacks interpersonal and leadership skills. He is responsible to help Rohnik in the testing phase of the SBPD project. He often approaches Glen for help in dealing with various issues about personnel and projects.

Neel is another major client of IH. Rohnik and Neel went to school together and still meet once a month for lunch. Rohnik has found Neel helpful in understanding how the political dynamics work at IH and how to deal with senior management members and decision-makers at IH.

Daniel has worked at ISSL for the last eight years as a senior software engineer in Glen's department. He has strong technical knowledge, and he is a good friend of Alan's. He is ambitious, and he wants to become a project manager. He couldn't get the position because of his lack of people skills. He is reluctant to share information with others and likes to keep his technical knowledge to himself to show his power. He is also a poor communicator and trainer.

Daniel has been assigned to Rohnik for a small part of a software design work package on a matrix basis. He has not completed his tasks on the SBPD project on time and shows a low level of commitment. Rohnik has tried to find answers for important questions about design and schedule, but he can never get a straight answer from Daniel.

Daniel is often late in meeting deadlines for tasks assigned to him on the SBPD project, and he does not come prepared in project meetings. He disrupts Rohnik in project meetings, challenges team decisions, and fails to work as a team. Rohnik is sometimes frustrated and feels like escalating this issue to Glen and Nikhil, but he believes Glen is protective of both Alan and Daniel, and he would not want to help him. Therefore, he tried to resolve problems himself with few results.

Issues: The project was going well. Glen suggested major changes the client might desire to the Steering Committee. He knows well that they will increase project scope and pose additional challenges for Rohnik to complete the deliverables within schedule and budget constraints.

Questions related to Part III of the book:

1. Identify the political positions taken by Rohnik, Nikhil, Glen, Alan, and Daniel.

2. What are the main political behaviors of Rohnik, Nikhil, Glen, Alan, and Daniel?

3. How should Rohnik deal with Glen and Daniel in the long term and in the short term?

Questions related to Part IV of the book:

4. Identify the main political issues at the management level in the following three categories and how Rohnik should manage these issues:

   - Organizational issues

   - Leadership issues

   - Project management issues

5. Identify the main political issues at the project level in the following three categories and how Rohnik should manage these issues:

   - Political issues and managing upward

   - Project management and team leadership issues

   - Stakeholder management issues

The Art of Positive Politics

End Notes

FOREWORD

1. Robert Block, *The Politics of Projects* (Yourdon Press, 1983).

2. Lee Bolman and Terrence Deal, *Reframing Organizations: Artistry, Choice, and Leadership,* 5th ed. (San Francisco: Jossey-Bass, 2013).

3. Vijay K. Verma, *Organizing Projects for Success: Volume 1 of the Human Aspects of Project Management* (Project Management Institute, 1995).

4. Vijay K. Verma, *Human Resource Skills for the Project Manager: Volume 2 of the Human Aspects of Project Management* (Project Management Institute, 1996).

5. Vijay K. Verma, *Managing the Project Team: Volume 3 of the Human Aspects of Project Management* (Project Management Institute, 1997).

PREFACE

1. Vijay K. Verma, *Human Resource Skills for the Project Manager*.

2. *A Guide to the Project Management Body of Knowledge (PMBOK® Guide)*, 5th ed. (Newtown Square, PA: Project Management Institute, 2013).

PART I

1. *A Guide to the Project Management Body of Knowledge (PMBOK® Guide)*, 5th ed.

CHAPTER 1

1. Vijay K. Verma, seminars presented at various locations on Leadership, Power, Influence, Politics, and Negotiations in Project Management, in Seminars World Program of Project Management Institute (PMI®) (2008–2015).

2. Jack Ferrarro, "My Project Advisor Newsletter," private communication, May 2010.

3. Ibid.

CHAPTER 2

1. Vijay K. Verma, seminars (2008–2015).

PART II

CHAPTER 3

1. Jeffrey K. Pinto, *Successful Information System Implementation: The Human Side* (PMI, 1994), 111–130.

2. Don R. Beeman and Thomas W. Sharkey, "The Use and Abuse of Corporate Politics," *Business Horizons* 36 (1987): 2, 26–30; M. L. Markus, "Power, Politics and MIS Implementation," *Communications of the ACM* 19 (1983): 321–342; Robert W. Allen, Dan L. Madison, Lyman W. Porter, Patricia A. Renwick, Bronston T. Mayes, "Organizational Politics: Tactics and Characteristics of Its Actors," *California Management Review* 22 (1979): 1, 78–93.

3. Jeffrey Pfeffer, *Power in Organizations* (Marshfield, MA: Pitman, 1981), p. 7.

4. Henry Mintzberg, *Power in and Around Organizations* (Englewood Cliffs, NJ: Prentice-Hall, 1983), p. 421.

5. Bronston T. Mayes and Robert W. Allen, "Toward a Definition of Organizational Politics," *Academy of Management Review* 2 (1977): 675.

6. Pinto, *Successful Information System Implementation*, pp. 111–130.

7. Verma, *Managing the Project Team*, pp. 133–148.

8. Geert Hofstede, *Cultures and Organizations: Software of the Mind* (New York: McGraw-Hill, 1993).

9. M. Dean Martin, "The Negotiation Differential for International Project Management." *Proceedings of the Annual Seminar/Symposium of the Project Management Institute* (1981).

10. Verma, *Managing the Project Team*, pp. 89–100.

11. Ibid.

12. Geert Hofstede, *Cultures and Organizations*, p. 14; *Cultural Dimensions in People Management: The Socialization Perspective in Globalizing Management, Creating and Learning the Competitive Organization* (New York: John Wiley, 1993), p. 14; and Charles H. Turner and Alfons Trompenaars, *The Seven Cultures of Capitalism: Value Systems for Creating Wealth in the United States, Japan, Germany, France, Britain, Sweden, and the Netherlands* (New York: Doubleday, 1993).

13. Stephen D. Owens and James Reagan McLaurin, "Cultural Diversity and Projects: What the Project Manager Needs to Know," *Proceedings of PMI93 Annual Seminar Symposium* (1993): 229–236.

CHAPTER 4

1. F. P. Doyle, "People-Power: The Global Human Resource Challenge for the '90s," *Columbia Journal and World Business* (1990): 36–45; L. R. Offerman and M. Gowing, "Organizations of the Future: Changes and Challenges," *American Psychologist* 45 (1990): 95–108.

2. Based on Rosabeth Moss Kanter, "Transcending Business Boundaries: 12,000 World Managers View Change," *Harvard Business Review* (1991): 151–164.

3. Verma, *Managing the Project Team*, pp. 18–20.

4. For additional perspectives on resistance to change, see Chris Argyris, *Overcoming Organizational Defenses: Facilitating Organizational Learning* (Boston: Allyn and Bacon, 1990); B. Spector, "From Bogged Down to Fired Up: Inspiring Organizational Change," *Sloan Management Review* (1989): 29–34; Chris Argyris, "Reasoning, Action Strategies, and Defensive Routines: The Case of OD Practitioners," in William A. Pasmore and Richard Woodman, eds. *Research in Organizational Change and Development 1* (Greenwich, CT: JAI Press, 1987), pp. 89–128; B. M. Staw, "Counterforces to Change," in Paul S. Goodman, ed. *Change in Organizations: New Perspectives on Theory and Research and Practice* (San Francisco: Jossey-Bass, 1982), pp. 87–121.

5. Based on Don Hellriegel, John W. Slocum, Jr., and Richard W. Woodman, *Organizational Behavior,* 6th ed. (St. Paul, MN: West Publishing Company, 1992), pp. 719–742.

6. Ibid; This strategy for minimizing the resistance to change is based on "How Companies Overcome Resistance to Change," *Management Review* 61, no. 11 (1972): 17–25.

7. Kurt Lewin, *Field Theory in Social Science* (New York: Harper & Row, 1951); and Kurt Lewin, "Frontiers in Group Dynamics," *Human Relations* 1 (1951): 5–41.

CHAPTER 5

1. *PMBOK® Guide*, pp. 391–397.

2. Ibid.

3. "Understanding Change Legacy and Tracking Risks," *Changefirst*, 2014, https://www.change-first.com/power-data-report-1-understanding-change-legacy-and-tracking-risks/.

4. Lynda Bourne and Derek H. T. Walker. "Advancing Project Management in Learning Organizations," *The Learning Organization* 11, no. 3 (2004): 226–243.

5. Lynn Crawford and V. Da Ros, "Politics and the Project Manager," *Australian Project Manager* 22, no. 4 (2002): 20.1–20.10.

6. Jeffrey K. Pinto, "Understanding the Role of Politics in Successful Project Management," *International Journal of Project Management*, 18 (2000): 85–91.

7. "Pulse of the Profession® In-Depth Report: Executive Sponsor Engagement: Top Driver of Project and Program Success," *PMI.org*, 2014, http://www.pmi.org/-/media/pmi/documents/public/pdf/learning/thought-leadership/pulse/executive-sponsor-engagement.pdf; "Managing Change in Organizations: A Practice Guide," *PMI.org*, 2013, https://www.pmi.org/pmbok-guide-standards/practice-guides/change.

8. H. James Harrington and Douglas Nelson, *The Sponsor as the Face of Organizational Change* (Newtown Square, PA: Project Management Institute, 2013).

9. David Miller and Mike Oliver, "PMI White Paper: Engaging Stakeholders for Project Success," Global Business Development, 2015, pp. 1–23, http://www.gbd.dk/files/984_engaging-stakeholders.pdf.

10. Verma, *Organizing Projects for Success*, pp. 46–70.

11. Tres Roeder, *Managing Project Stakeholders: Building a Foundation to Achieve Project Goals* (Hoboken, NJ: John Wiley and Sons, Inc., 2013), pp. 97–101.

12. Adrienne Watt, *Project Management, BC Open Textbooks*, 2010, Chapter 5, pp.1–10, https://opentextbc.ca/projectmanagement/.

13. Roeder, *Managing Project Stakeholders*, pp. 31–38; and *PMBOK® Guide*, pp. 395–397.

14. Ibid.

15. Robert Newcombe, "From Client to Project Stakeholder: A Stakeholder Mapping Approach," *Construction Management and Economics* 21, no. 8 (2003): 841–848.

16. Miller and Oliver, "Engaging Stakeholders for Project Success."

17. Crawford and Da Ros, "Politics and the Project Manager."

PART III

CHAPTER 6

1. Verma, seminars (2014); Jeffrey K. Pinto, *Power & Politics in Project Management* (Newtown Square, PA: Project Management Institute, 1998), pp. 75–77; and Pinto, *Successful Information System Implementation*, pp. 118–122.

2. Wikipedia contributors, "Flounder," *Wikipedia, The Free Encyclopedia,* https://en.wikipedia.org/w/index.php?title=Flounder&oldid=797660753.

3. Verma, seminars (2014); Pinto, *Power & Politics in Project Management*, pp. 75–77; and Pinto, *Successful Information System Implementation*, pp. 118–122.

4. Dale Myers, "Corporate Politics for Project Managers 101." *Dale Myers' Blog: Think. Plan. Act. Repeat.* (blog), July 09, 2012, https://dalemyers.wordpress.com/2012/07/09/corporate-politics-for-project-managers-101/.

5. Wikipedia contributors, "Shark," *Wikipedia, The Free Encyclopedia,* https://en.wikipedia.org/w/index.php?title=Shark&oldid=800170071.

6. Verma, seminars (2014); Pinto, *Successful Information System Implementation*, pp. 118–122; and Pinto, *Power & Politics in Project Management*, pp. 75–77.

7. Verma, seminars (2014); and Myers, "Corporate Politics for Project Managers 101."

8. Harvey V. McKay, *Swimming with the Sharks Without Being Eaten Alive* (New York: Harper Collins, 2005), pp. ix–xviii.

9. Ibid.

10. Wikipedia contributors, "Dolphin," *Wikipedia, The Free Encyclopedia,* https://en.wikipedia. org/w/index.php?title=Dolphin&old id=800221291.

11. Verma, seminars (2014); Pinto, *Power & Politics in Project Management,* pp. 75–77; and Pinto, *Successful Information System Implementation,* pp. 118–122.

12. Verma, seminars (2014); Myers, "Corporate Politics for Project Managers 101"; and Pinto, *Power & Politics in Project Management,* pp. 146–152.

13. Bernard Keys and Thomas Case, "How to Become an Influential Manager," *Academy of Management Executive,* 4, no. 4 (1990): 38–51.

14. Roger Fisher and William L. Ury, *Getting to Yes: Negotiating Agreement without Giving In* (New York: Houghton Mifflin, 1981).

15. Jeffrey K. Pinto and Om Prakash Kharbanda, *Successful Project Managers: Leading Your Team to Success (Industrial Engineering)* (New York: Van Nostrand Reinhold, 1995).

16. Sun Tzu, *The Art of War: The Oldest Military Treatise in the World* (Dover Publications, Inc., 2002). This is an unabridged republication of an edition originally published in 1944 by the Military Service Publishing Company, Harrisburg, Pennsylvania, USA, translated by Lionel Giles, Assistant, Department of Oriental Printed Books and Manuscripts.

17. Cynthia Hardy, *Strategies for Retrenchment and Turnaround: The Politics of Survival* (Berlin: Walter de Gruyter, 1990), Chapter 14; S. C. Goh and A. R. Doucet, "Antecedent Situational Conditions of Organizational Politics: An Empirical Investigation," *Proceedings of the Annual ASAC Conference, Organizational Behavior Division 7,* pt. 5 (1986): 77–86; Todd D. Jick and Victor V. Murray, "The Management of Hard Times: Budget Cutbacks in Public Sector Organizations," *Organization Studies* 3 (1982): 141–69; and Jeffrey Gandz and Victor V. Murray, "The Experience of Workplace Politics," *Academy of Management Journal* 23 (1980): 237–51.

18. G. R. Ferris, G. S. Russ, and P. M. Fandt, "Politics in Organizations" in *Impression Management in the Organization,* eds. Robert A. Giacalone and Paul Rosenfeld (Hillsdale, NJ: Erlbaum, 1989), pp. 143–70; and H. Mintzberg, "The Organization as Political Arena," *Journal of Management Studies* 22 (1985): 133–54.

19. Lyman W. Porter, Robert W. Allen, and Harold L. Angle, "The Politics of Upward Influence in Organizations," *Research in Organizational Behavior* 3, (1981): 120–22; and Robert J. House, "Power and Personality in Complex Organizations," *Research in Organizational Behavior* 10 (1988): 305–57.

20. Peter E. Mudrack, "An Investigation into the Acceptability of Workplace Behaviors of a Dubious Ethical Nature," *Journal of Business Ethics* 12, (1993): 517–24; and Richard Christie, *Studies in Machiavellianism* (New York: Academic Press, 1970).

21. McKay, *Swimming with the Sharks.*

22. Myers, "Corporate Politics for Project Managers 101"; and Pinto, *Power & Politics in Project Management.*

CHAPTER 7

1. Doyle, "People-Power," pp. 36–45; and Offermann and Gowing, "Organizations of the Future," pp. 95–108.

2. Sunny Baker and Kim Baker, *On Time/On Budget: A Step-By-Step Guide for Managing Any Project* (Englewood Cliffs, NJ: Prentice Hall, 1992), 211–220; and Warren G. Bennis, Kenneth D. Benne, and Robert Chin, ed. *The Planning of Change: Readings in the Applied Behavioral Sciences* (New York: Holt, Rinehart & Winston, 1961), 69.

3. Verma, seminars (2014); Steven L. McShane, Sandra L. Steen, and Kevin Tasa, *Canadian Organizational Behaviour 9* (Ontario: McGraw-Hill Ryerson, 2014); Steven L. McShane, *Canadian Organizational Behaviour, Second Edition* (Chicago: Richard D. Irwin, Inc., 1995), 355–375; Victor V. Murray and Jeffrey Gandz, "Games Executives Play: Politics at Work," *Business Horizons* (1980): 11–23; and Allen, Madison, Porter, Renwick & Mayes, "Organizational Politics," pp. 77–88.

4. Blake E. Ashforth and Raymond T. Lee, "Defensive Behavior in Organizations: A Preliminary Model," *Human Relations* 43, (1990): 621–48.

5. David Krackhardt and Jeffrey R. Hanson, "Informal Networks: The Company Behind the Chart," *Harvard Business Review* 71, (1993): 104–111; and Robert E. Kaplan, "Trade Routes: The Manager's Network of Relationships," *Organizational Dynamics* (1984): 37–52.

6. Elizabeth A. Mannix, "Organizations as Resource Dilemmas: The Effects of Power Balance on Coalition Formation in Small Groups," *Organizational Behavior and Human Decision Process* 55 (1993): 1–22; Anthony T. Cobb, "Toward the Study of Organizational Coalitions: Participant Concerns and Activities in a Simulated Organizational Setting," *Human Relations* 44 (1991): 1057–1079; and William B. Stevenson, Jone L. Pearce, and Lyman W. Porter, "The Concept of 'Coalition' in Organization Theory and Research," *Academy of Management Review* 10 (1985): 256–268.

7. Ronald J. Burke and Carol A. McKeen, "Women in Management," *International Review of Industrial and Organizational Psychology* 7 (1992): 245–283; and B. R. Ragins and E. Sundstrom, "Gender and Power in Organizations: A Longitudinal Perspective," *Psychological Bulletin* 105, (1989): 51–88.

8. Allan R. Cohen and David L. Bradford, "Influence Without Authority: The Use of Alliances, Reciprocity and Exchange to Accomplish Work," *Organizational Dynamics* 17, no. 3 (1989): 5–17.

9. Robert A. Giacalone and Paul Rosenfeld, eds., *Applied Impression Management* (Newbury Park, CA: SAGE Publications, Inc., 1991; and James T. Tedeschi, ed., *Impression Management Theory and Social Psychological Research* (New York: Academic Press, 1981).

10. William L. Gardner III, "Lessons in Organizational Dramaturgy: The Art of Impression Management,", *Organizational Dynamics* (1992): 33–46; Robert C. Liden and Terence R. Mitchell, "Ingratiatory Behaviors in Organizational Settings," *Academy of Management Review* 13 (1988): 572–587; and A. MacGillivray, S. Ashcroft, and M. Stebbins, "Meritless Ingratiation," *Proceedings of the Annual ASAC Conference, Organizational Behavior Division* 7, pt. 7 (1986): 127–135.

11. Paul G. Stern and Tom Shachtman, *Straight to the Top: Beyond Loyalty, Gamesmanship, Mentors, and Other Corporate Myths* (Grand Central Pub, 1990), 20–21.

CHAPTER 8

1. Verma, seminars (2014); McShane, *Canadian Organizational Behaviour 9*; and McShane, *Canadian Organizational Behaviour, Second Edition*, pp. 355–375.

2. Ibid; and Verma, *Managing the Project Team*, pp. 89–110.

3. Verma, seminars (2014); McShane, *Canadian Organizational Behaviour 9*; and McShane, *Canadian Organizational Behaviour, Second Edition*, pp. 355–375.

4. Hofstede, *Cultures and Organizations*, p. 5; and Martin, "The Negotiation Differential for International Project Management."

5. Vern Terpstra, *The Cultural Environment of International Business* (Dallas, TX: South Western Publishing, 1978), 2, 176; and Vern Terpstra, *International Marketing* (Hinsdale, IL: Dryden, 1972), 83.

6. Hofstede, *Cultures and Organizations*, p.5.

7. Fred Luthans and Richard Hodgetts, *International Management* (New York: McGraw-Hill, 1991), 35.

8. Verma, seminars (2014); Verma, *Managing the Project Team*, pp. 89–110; and Martin, "The Negotiation Differential for International Project Management."

9. Hofstede, *Cultures and Organizations*, p. 14, *Cultural Dimensions in People Management*, p. 14; Turner and Trompenaars, *The Seven Cultures of Capitalism*; Stephen D. Owens and James Reagan McLaurin, "Cultural Diversity and Projects," pp. 229–236.

10. Richard A. Punzo, "Managing Cross Cultural Values in Project Teams," *Proceedings of Annual Seminar/Symposium* (Upper Darby, PA: Project Management Institute, 1996), 863–868; and Training Management Corporation, *Effective Global Manager,* Seminar and Workbook, Princeton, NJ (1996).

11. Edward T. Hall, *The Silent Language* (New York: Doubleday Books, 1990); and Florence Rockwood Kluckhohn, *Variations in Value Orientations* (Evanston, IL: Row, Peterson, 1961).

12. Ibid.

13. Ibid.

14. Punzo, "Managing Cross Cultural Values in Project Teams"; Stephen Rhinesmith, *A Manager's Guide in Globalization: Six Keys to Success in a Changing World* (Homewood, IL: Business One Irwin Press, 1993); and Hall, *The Silent Language.*

15. Ibid; and Edward C. Stewart and Milton J. Bennett, *American Cultural Patterns: A Cross-Cultural Perspective* (Yarmouth, ME: Intercultural Press, 1991).

16. Daniel J. Brass and Marlene E. Burkhardt, "Potential Power and Power Use: An Investigation of Structure and Behavior," *Academy of Management Journal* 36, no. 3 (1993): 441–470, 1993; Judith Dozier Hackman, "Power and Centrality in the Allocation of Resources in Colleges and Universities," *Administrative Science Quarterly* 30 (1985): 61-77.David J. Hickson, C. R. Hinings, C. A. Lee, R. E. Schneck, and J. M. Pennings, "A Strategic Contingencies Theory of Intraorganizational Power," *Administrative Science Quarterly* 16, no. 2 (1971): 219–221.

17. Bob Nelson, *1001 Ways to Energize Employees,* (New York: Workman Publishing Company, 1997).

PART IV

CHAPTER 9

1. Stephen R. Covey, *Seven Habits of Highly Effective People* (New York: Simon and Schuster, 2000).

2. Katie Carlone and Linda A. Hill, *Managing Up: Expert Solutions to Everyday Challenges* (Boston: Harvard Business School Publishing Corporation, 2008): v–vii, 4–7.

3. Deborah Singer Dobson and Michael Singer, *Managing Up: 59 Ways to Build a Career-Advancing Relationship with Your Boss* (New York: AMACOM, 2000).

CHAPTER 10

1. Verma, seminars (2012–2015), *Human Resource Skills for Project Managers*, pp. 239–242.

2. Richard S. Wellins, William C. Byham, and Jeanne M. Wilson, *Empowered Teams* (San Francisco: Jossey-Bass Inc., 1991), 21–23; and Cynthia D. Scott and Dennis T. Jaffe, *Empowerment* (Menlo Park, CA: Crisp Publications, Inc., 1991), i–ii, 4–16, 29–37).

3. James M. Kouzes and Barry Z. Posner, *The Leadership Challenge*, 4th ed. (San Francisco: John Wiley and Sons Inc., 2007), 3–27.

CHAPTER 11

1. Verma, seminars (2012–2015), *Human Resource Skills for Project Managers*, pp. 239–242.

2. PMBOK® Guide, pp. 391–392.

3. Verma, *Organizing Projects for Success*, pp. 46–70.

4. Roeder, *Managing Project Stakeholders*, pp. 31–37.

5. Roeder, *Managing Project Stakeholders*, pp. 31–37; Omar Muhammad, "Complex Projects and the Politics of Stakeholder Management," *PMtimes: Resources for Project Managers*, 2013, https://www.projecttimes.com/articles/complex-projects-and-the-politics-stakeholder-management.

6. Omar Muhammad, "Complex Projects and the Politics of Stakeholder Management."

7. Roeder, *Managing Project Stakeholders*, pp. 31–37.

8. Ibid.

9. Tres Roeder, *A Sixth Sense for Project Management* (Bloomington, IN: AuthorHouse, 2011).

10. Ibid.

11. Peter Block, *The Empowered Manager: Positive Political Skills at Work* (San Francisco: John Wiley and Sons, Inc., 1987), 137–160.

12. Ibid.

13. Verma, seminars (2012–2015); Block, *The Empowered Manager*, pp. 137–160.

14. Verma, seminars (2012–2015).

15. Verma, *Organizing Projects for Success*, pp. 99–102, 119–120.

16. Verma, *Managing the Project Team*, pp. 122–123, 216.

17. Verma, *Human Resource Skills for the Project Manager*, pp. 24–25, 142–143.

18. Ibid.

References

A Guide to the Project Management Body of Knowledge (PMBOK Guide). 5th. Newtown Square, PA: Project Management Institute, 2013.

Allen, Robert W., Dan L. Madison, Lyman W. Porter, Patricia A. Renwick, and Bronston T. Mayes. "Organizational Politics: Tactics and Characteristics of Its Actors." *California Management Review* 22 (1979).

Argyris, Chris. *Overcoming Organizational Defenses: Facilitating Organizational Learning.* Boston: Allyn and Bacon, 1990.

Argyris, Chris. "Reasoning, Action Strategies, and Defensive Routines: The Case of OD Practitioners." In *Research in Organizational Change and Development 1*, edited by William A. Pasmore and Richard Woodman. Greenwich, CT: JAI Press, 1987.

Ashforth, Blake E., and Raymond T. Lee. "Defensive Behavior in Organizations: A Preliminary Model." *Human Relations* 43 (1990).

Baker, Sunny, and Kim Baker. *On Time/On Budget: A Step-By-Step Guide for Managing Any Project*. Englewood Cliffs, NJ: Prentice Hall, 1992.

Beeman, Don R., and Thomas W. Sharkey. "The Use and Abuse of Corporate Politics." *Business Horizons* 36 (1987).

Bennis, Warren G., Kenneth D. Benne, and Robert Chin, . *The Planning of Change: Readings in the Applied Behavioral Sciences*. New York: Holt, Rinehart & Winston, 1961.

Block, Peter. *The Empowered Manager: Positive Political Skills at Work*. San Francisco: John Wiley and Sons, 1987.

Block, Robert. *The Politics of Projects*. Yourdon Press, 1983.

Bolman, Lee, and Terrence Deal. *Reframing Organizations: Artistry, Choice, and Leadership*. 5th. San Francisco: Jossey-Bass, 2013.

References

Bourne, Lynda, and Derek H. T. Walker. "Advancing Project Management in Learning Organizations." *The Learning Organization* 11, no. 3 (2004).

Brass, Daniel J., and Marlene E. Burkhardt. "Potential Power and Power Use: An Investigation of Structure and Behavior." *Academy of Management Journal* 36, no. 3 (1993).

Burke, Ronald J., and Carol A. McKeen. "Women in Management." *International Review of Industrial and Organizational Psychology* 7 (1992).

Carlone, Katie, and Linda A. Hill. *Managing Up: Expert Solutions to Everyday Challenges.* Boston: Harvard Business School Publishing Corporation, 2008.

Christie, Richard. *Studies in Machiavellianism.* New York: Academic Press, 1970.

Cobb, Anthony T. "Toward the Study of Organizational Coalitions: Participant Concerns and Activities in a Simulated Organizational Setting." *Human Relations* 44 (1991).

Cohen, Allan R., and David L. Bradford. "Influence Without Authority: The Use of Alliances, Reciprocity and Exchange to Accomplish Work." *Organizational Dynamics* 17, no. 3 (1989).

Covey, Stephen r. *Seven Habits of Highly Effective People.* New York: Simon and Schuster, 2000.

Crawford, Lynn, and V. Da Ros. "Politics and the Project Manager." *Australian Project Manager* 22, no. 4 (2002).

Dobson, Deborah Singer, and Michael Singer. *Managing Up: 59 Ways to Build a Career-Advancing Relationship with Your Boss.* New York: AMACOM, 2000.

Doyle, F. P. "People-Power: The Global Human Resource Challenge for the '90s." *Columbia Journal and World Business,* 1990.

Ferrarro. "My Project Advisor Newsletter." private communication, May 2010.

Ferris, G. R., G. S. Russ, and P. M. Fandt. "Politics in Organizations." In *Impression Management in the Organization,* edited by Robert A. Giacalone and Paul Rosenfeld. Hillsdale, NJ: Erlbaum, 1989.

Fisher, Roger, and William L. Ury. *Getting to Yes: Negotiating without Giving In.* New York: Houghton Mifflin, 1981.

Gandz, Jeffrey, and Victor V. Murray. "The Experience of Workplace Politics." *Academy of Management Journal* 23 (1980).

Gardner, III, William L. "Lessons in Organizational Dramaturgy: The Art of Impression Management." *Organizational Dynamics,* 1992.

Giacalone, Robert A., and Paul Rosenfeld, . *Applied Impression Management*. Newbury Park, CA: SAGE Publications, Inc., 1991.

Goh, S. C., and A. R. Doucet. "Antecedent Situational Conditions of Organizational Politics: An Empirical Investigation." *Proceedings of the Annual ASAC Conference, Organizational Behavior Division* 7. 1986.

Hackman, Judith Dozier. "Power and Centrality in the Allocation of Resources in Colleges and Universities." *Administrative Science Quarterly* 30 (1985).

Hall, Edward T. *The Silent Language*. New York: Doubleday Books, 1990.

Hardy, Cynthia. *Strategies for Retrenchment and Turnaround: The Politics of Survival*. Berlin: Walter de Gruyter, 1990.

Harrington, H. James, and Douglas Nelson. *The Sponsor as the Face of Organizational Change*. Newtown Square, PA: Project Management Institute, 2013.

Hellriegel, Don, Jr., John W. Slocum, and Richard W. Woodman. *Organizational Behavior*. 6th. St. Paul, MN: West Publishing Company, 1992.

Hickson, David J., C. R. Hinings, C. A. Lee, R. E. Schneck, and J. M. Pennings. "A Strategic Contingencies Theory of Intraorganizational Power." *Administrative Science Quarterly* 16, no. 2 (1971).

Hofstede, Geert. *Cultural Dimensions in People Management: The Socialization Perspective in Globalizing Management, Creating and Learning the Competitive Organization.* New York: John Wiley, 1993.

—. *Cultures and Organizations: Software of the Mind.* New York: McGraw-Hill, 1993.

House, Robert J. "Power and Personality in Complex Organizations." *Research in Organizational Behavior* 10 (1988).

"How Companies Overcome Resistance to Change." *Management Review* 61, no. 11 (1972).

Jick, Todd D., and Victor V. Murray. "The Management of Hard Times: Budget Cutbacks in Public Sector Organizations." *Organization Studies* 3 (1982).

Kanter, Rosabeth Moss. "Transcending Business Boundaries: 12,000 World Managers View Change." *Harvard Business Review*, 1991.

Kaplan, Robert E. "Trade Routes: The Manager's Network of Relationships." *Organizational Dynamics*, 1984.

Keys, Bernard, and Thomas Case. "How to Become an Influential Manager." *Academy of Management Executive* 4, no. 4 (1990).

Kluckhohn, Florence Rockwood. *Variations in Value Orientations.* Evanston, IL: Row, Peterson, 1961.

Krackhardt, David, and Jeffrey R. Hanson. "Informal Networks: The Company Behind the Chart." *Harvard Business Review* 71 (1993).

Lewin, Kurt. *Field Theory in Social Science.* New York: Harper & Row, 1951.

Lewin, Kurt. "Frontiers in Group Dynamics." *Human Relations* 1 (1951).

Liden, Robert C., and Terence R. Mitchell. "Ingratiatory Behaviors in Organizational Settings." *Academy of Management Review* 13 (1988).

Luthans, Fred, and Richard Hodgetts. *International Management.* New York: McGraw-Hill, 1991.

MacGillivray, A., S. Ashcroft, and M. Stebbins. "Meritless Ingratiation." *Proceedings of the Annual ASAC Conference, Organizational Behavior Division.* 1986.

"Managing Change in Organizations: A Practice Guide." *PMI.org.* 2013. https://www.pmi.org/pmbok-guide-standards/practice-guides/change.

Mannix, Elizabeth A. "Organizations as Resource Dilemmas: The Effects of Power Balance on Coalition Formation in Small Groups." *Organizational Behavior and Human Decision Process* 55 (1993).

Markus, M. L. "Power, Politics and MIS Implementation." *Communications of the ACM* 19 (1983).

Martin, M. Dean. "The Negotiation Differential for International Project Management." *Proceedings of the Annual Seminar/Symposium of the Project Management Institute.* 1981.

Mayes, Bronston T., and Robert W. Allen. "Toward a Definition of Organizational Politics." *Academy of Management Review* 2 (1977).

McKay, Harvey. *Swimming with the Sharks Without Being Eaten Alive.* New York: Harper Collins, 2005.

McShane, Steven L. *Canadian Organizational Behaviour, Second Edition.* Chicago: Richard D. Irwin, Inc., 1995.

McShane, Steven L., Sandra L. Steen, and Kevin Tasa. *Canadian Organizational Behaviour 9.* Ontario: McGraw-Hill Ryerson, 2014.

Miller, David, and Mike Oliver. "PMI White Paper: Engaging Stakeholders for Project Success." *Global Business Development.* 2015. http://www.gbd.dk/files/984_engagingstakeholders.pdf.

Mintzberg, Henry. *Power in and Around Organizations.* Englewood Cliffs, NJ: Prentice-Hall, 1983.

Mintzberg, Henry. "The Organization as Political Arena." *Journal of Management Studies* 22 (1985).

Mudrack, Peter E. "An Investigation into the Acceptability of Workplace Behaviors of a Dubious Ethical Nature." *Journal of Business Ethics* 12 (1993).

Murray, Victor V., and Jeffrey Gandz. "Games Executives Play: Politics at Work." *Business Horizons,* 1980.

Myers, Dale. "Corporate Politics for Project Managers 101." *Dale Myers' Blog: Think. Plan. Act. Repeat.* July 09, 2012. https://dalemyers. wordpress.com/2012/07/09/corporate-politics-for-project-managers-101/.

Nelson, Bob. *1001 Ways to Energize Employees.* New York: Workman Publishing Company, 1997.

Newcombe, Robert. "From Client to Project Stakeholder: A Stakeholder Mapping Approach." *Construction Management and Economics* 21, no. 8 (2003).

Offerman, L. R., and M. Gowing. "Organizations of the Future: Changes and Challenges." *American Psychologist* 45 (1990).

Owens, Stephen D., and James Reagan McLaurin. "Cultural Diversity and Projects: What the Project Manager Needs to Know." *Proceedings of PMI93 Annual Seminar Symposium.* 1993.

Pfeffer, Jeffrey. *Power in Organizations.* Marshfield, MA: Pitman, 1981.

Pinto, Jeffrey K. *Power & Poliltics in Project Management*. Newtown Square, PA: Project Management Institute, 1998.

Pinto, Jeffrey K. *Successful Information System Implementation: The Human Side*. PMI, 1994.

Pinto, Jeffrey K. "Understanding the Role of Politics in Successful Project Management." *International Journal of Project Management* 18 (2000).

Pinto, Jeffrey K., and Om Prakash Kharbanda. *Successful Project Managers: Leading Your Team to Success (Industrial Engineering)*. New York: Van Nostrand Reinhold, 1995.

Porter, Lyman W., Robert W. Allen, and Harold L. Angle. "The Politics of Upward Influence in Organizations." *Research in Organizational Behavior* 3 (1981).

"Pulse of the Profession In-Depth Report: Executive Sponsor Engagement: Top Driver of Project and Program Success." *PMI.org*. 2014. http://www.pmi.org/-/media/pmi/documents/public/pdf/learning/thought-leadership/pulse/executive-sponsor-engagement.pdf.

Punzo, Richard A. "Managing Cross Cultural Values in Project Teams." *Proceedings of Annual Seminar/Symposium*. Upper Darby: Project Management Institute, 1996.

Ragins, B. R., and E. Sundstrom. "Gender and Power in Organizations: A Longitudinal Perspective." *Psychological Bulletin* 105 (1989).

Rhinesmith, Stephen. *A Manager's Guide in Globalization: Six Keys to Success in a Changing World.* Homewood, IL: Business One Irwin Press, 1993.

Roeder, Tres. *A Sixth Sense for Project Management.* Bloomington, IN: AuthorHouse, 2011.

—. *Managing Project Stakeholders: Building a Foundation to Achieve Project Goals.* Hoboken, NJ: John Wiley and Sons, Inc., 2013.

Scott, Cynthia D., and Dennis T. Jaffe. *Empowerment.* Menlo Park, CA: Crisp Publications, Inc., 1991.

Spector, B. "From Bogged Down to Fired Up: Inspiring Organizational Change." *Sloan Management Review,* 1989.

Staw, B. M. "Counterforces to Change." In *Change in Organizations: New Perspectives on Theory and Research and Practice,* edited by Paul S. Goodman. San Francisco: Jossey-Bass, 1982.

Stern, Paul G., and Tom Shachtman. *Straight to the Top: Beyond Loyalty, Gamesmanship, Mentors, and Other Corporate Myths.* Grand Central Pub, 1990.

Stevenson, William B., Jone L. Pearce, and Lyman W. Porter. "The Concept of "Coalition" in Organization Theory and Research." *Academy of Management Review* 10 (1985).

Stewart, Edward C., and Milton J. Bennett. *American Cultural Patterns: A Cross-Cultural Perspective.* Yarmouth, ME: Intercultural Press, 1991.

Tedeschi, James T., ed. *Impression Management Theory and Social Psychological Research.* New York: Academic Press, 1981.

Training Management Corporation. *Effective Global Manager.* Princeton, NJ, 1996.

Turner, Charles H., and Alfons Trompenaars. *The Seven Cultures of Capitalism: Value Systems for Creating Wealth in the United States, Japan, Germany, France, Britain, Sweden, and the Netherlands.* New York: Doubleday, 1993.

Tzu, Sun. *The Art of War: The Oldest Military Treatise in the World.* Dover Publications, Inc., 2002.

Understanding Change Legacy and Tracking Risks. 2014. https://www.changefirst.com/power-data-report-1-understanding-change-legacy-and-tracking-risks/.

Verma, Vijay K. *Human Resource Skills for the Project Manager: Volume 2 of the Human Aspects of Project Management.* Project Management Institute, 1996.

—. *Managing the Project Team: Volume 3 of the Human Aspects of Project Management.* Project Management Institute, 1997.

—. *Organizing Projects for Success; Volume 1 of the Human Aspects of Project Management.* Project Management Institute, 1995.

—. "Seminars Presented at Various Locations on Leadership, Power, Influence, Politics, and Negotiations in Project Management, in Seminars World Program of Project Management Institute (PMI)." 2008-2015.

Watt, Adrienne. "Project Management." *BC Open Textbooks.* 2010. https://opentextbc.ca/project-management/.

Wellins, Richard S., William C. Byham, and Jeanne M. Wilson. *Empowered Teams.* San Francisco: Jossey-Bass Inc., 1991.

Wikipedia contributors. "Dolphin." *Wikipedia, The Free Encyclopedia.* n.d. https://en.wikipedia.org/w/index.php?title=Dolphin&oldid=800221291.

Wikipedia contributors. "Flounder." *Wikipedia, The Free Encyclopedia.* n.d. https://en.wikipedia.org/w/index.php?title=Flounder&oldid=797660753.

Wikipedia contributors. "Shark." *Wikipedia, The Free Encyclopedia.* n.d. https://en.wikipedia.org/w/index.php?title=Shark&oldid=800170071.

The Art of Positive Politics

Index

Page numbers in italics refer to figures and tables.

A

acceptance
  agreement and, 110–13, *113*, 260–61, 323–25
  gaining, *186*, 187, 260–61
accountability, deflecting, 98–99, *218*, 219–20,
  243–44
  See also finger-pointing
acquisitions, 136, 158
actions
  reducing risk of wrong, *177*, 178–79
  ensuring positive, *177*, 180

adversaries, 393–94, 398, 407, 420–21

agendas,
    competing, *99*, 100–101
    hidden, *99*, 100–101, *136*, 150, 172, 219

agents. *See* consultants, external

agreement
    acceptance and, 110–13, 260–61, 323–25
    trust and, *57*, 191–92, *386*, 393–98

Allen, Robert W., 96–97, 234, 242

alliances, *106*, 113–14, 390

allies, 418, 421
    keeping, *190*, 191, *193*, 194, *196*, 197
    managing, 393–95
    possible, *177*, 180–81, 184, *186*, 187

amnesty rule, *351*, 358–59

Angle, Harold L., 234

anxiety, 293–94

Ashcroft, S., 263

Ashforth, Blake E., 244

B

Baker, Kim, 241

Baker, Sunny, 241

bedfellows, *394*, 396

Beeman, Don R., 96

benefits
    personal, 98, *99*, 100, 150, *235*
    potential of change, 121, 156, 246
    project, *177*, 181, 193–94

Benne, Kenneth D., 241

Bennett, Milton J., 288

Bennis, Warren G., 241

blaming. *See* finger-pointing; search for the guilty

Block, Peter, 393–94

Block, Robert, 20

Bolman, Lee, 21

boredom, *136*, 152

Bourne, Lynda, 165

Bradford, David L., 259

brainstorming, *275*, 281–83

Brass, Daniel J., 295

Burke, Ronald J., 255

Burkhardt, Marlene E., 295

buy-in,
    commitment and, 111–13
    gaining, *106*, 108–13
    support and, *324*, 390–91
    See also acceptance; agreement; commitment

Byham, William C., 345

C

Carlone, Katie, 319

Case, Thomas, 229

celebration, 70, *71*, 73, *78*, 78–79

centrality, 295

champions, 139–46, 197, 346–47, 366–69, *369*

change
    barriers to 245
    business environment, 154, 159, 221–22

commitment to, 176, 187–90
effecting, or impact, 176, *177*, 178, 183–84, 389
introducing, *243*, 245–46
management of, *136*, 155–56
organizational, 153–54
resistance to, 155–56, 245–46
Changefirst, 164
Chin, Robert, 241
Christie, Richard, 237
churning, *251*, 256–57
coalitions, 130, *251*, 254–56
Cobb, Anthony T., 254
Cohen, Allan R., 259
collaboration, *99*, 102, *106*, 115–16, 225
commitment, 168, 262, 379–80
buy-in and, 108–13
to change, 176, 187–90
motivation and, *113*
versus compliance, 119–21
communication, *119*, 168, 248, 260, 287–88
problems, 245, *386*, 403–404
skills, *225*, 230–31, 319, 391–92
strategies, 176–98
system, 332–33
See also information
competition, *106*, 115–16
compliance, 119–21
conclusions, forming, *258*, 264–65
Concurrent Engineering, 121
conflict, *68*
Dolphins and, *225*, 231–32

management of, 231–32, *275*, 280–81
resolution of, *386*, 403–5

confusion, *71*, *386*, 401–2

consultants, *332*, 336–37, *365*, 370–72

control, 99–100, 236
command and, 54–55
information, 251–52
internal focus of, *235*, 236

cooperation, *106*, 113–14, *177*, 183, *225*, 229

Covey, Stephen R., 315

Crawford, Lynn, 165, 198

cultural differences. *See* culture, diversity of

culture
aspects of *136*, 155, *275*, 284–88
diversity of, *106*, 128–31, 286–87, 404
elements of, 129–30, 286
variables and associated orientations of thinking, 287–88

D

Da Ros, V. 165, 198

Deal, Terrence, 21

decision-making, 419–20
avoidance in, *258*, 266–67
inappropriate, *136*, 147
problem solving and, *275*, 283–84
subjective, *99*, 101, 138–39

disillusionment, 70, *71*, 73, *78*, 78–80

Dobson, Deborah Singer, 319

Doing things Right the First Time. *See* DRFT

Dolphins, 425–29
  dealing with politics, 212–13, 224–34
  management techniques, 227–34, *365*, 376–77
  political viewpoints, 224–25
Doucet, A. R., 234
Doyle, F. P., 155, 241
DRFT (Doing things Right the First Time), 120–21

E

egos and selfishness, 148–49
emotional quotient. *See* EQ
empowerment, *177*, 181–82, 261–62, *340*, 344–46
EQ (emotional quotient), 299, *340*, 343
ethics, 192–95
evaluation, *351*, 354–55
experience, *53*, 55–56, *177*, 178, 247

F

facilitation, *275*, 281–83
Fandt, P. M., 234
favoritism, 78, 82, *99*, 101–2, *106*, 151–52
feedback, 231, 246, 325,
  360-degree, *340*, 347–49
fence sitters, 394, 397, 420
Ferrarro, Jack, 55
Ferris, G. R., 234
finger-pointing, 71, 98–99, *218*, 220, 243–44
  See also search for the guilty

Fisher, Roger, 230

Flounders, 211–16, 226, *227*

Force Field Analysis, 156

functional silos, *99*, 102

G

Gandz, Jeffrey, 234, 242

Gardner, III, William L., 263

Giacalone, Robert A., 263

global market environment, 159
    issues, 38, 92, 94, 135, 158

goals
    clarifiying, 401–2
    individual performance, 335–36
    organizational, 106–13, 334–35
    stakeholder, 171–73

Goh, S. C., 234

Gowing, M., 155, 241

*Guide to the Project Management Body of Knowledge,
A.* See *PMBOK Guide*

H

Hackman, Judith Dozier, 295

Hall, Edward T., 287

Hanson, Jeffrey R., 247

Hardy, Cynthia, 234

Harrington, H. James, 167

Hellriegel, Don, 156

Hickson, David J., 295

Hill, Linda A., 319

Hillman, Sidney, 41

Hinings, C. R., 295

Hodgetts, 285

Hofstede, Geert, 129–30, 285–86

House, Robert J., 234

Human Resource Skills for the Project Manager, 22, 26

I

impact
    negative, 243–57, *258*, 263–68
    positive, *243*, 244–56, *258*, 259–64, 268–69
    versus power, *177*, 183–84

impressions, managing, *258*, 263–64

improvements, incremental, *251*, 253–54

influence, 49, 59–60, 419
    champions and, 140–42, *146*
    Dolphins and, *225*, 226–29
    negative, *177*, 179, *196*, 198
    political in a project environment, 68
    stakeholders and, 176, *177*, 181–83, 389
    versus power, 176, *177*, 181–83

information, 267, 404
    controlling, 99, *251*, 251–52
    encouraging sharing, *106*, 125–26, 275, 275–76, 278, 380
    filtering, *251*, 252–53, *386*, 405–6, 422–24
    informal power of, 57, 276
    sharing, *243*, 249–50

Sharks and, *218*, 220
stakeholder commitment to change and, *188*,
     188–89
stakeholder ethics and, *193*, 193–94
stakeholder impact and, *177*, 183–84
stakeholder influence and, *177*, 181–83
stakeholder integrity and, *196*, 196–97
stakeholder interest and, *177*, 180
stakeholder knowledge and, *177*, 178
stakeholder predictability and, *186*, 186
stakeholder respect and, *190*, 191
withholding, 125, *136*, 151, *235*, 236
initiatives
     interpersonal and team, 274–88
     organizational, 289–99
integrity, 195–98
interest
     best, 321–23
     champions and vested, 140
     stakeholder 171–73, 176–84, *186*, 187
     versus power, 179–81
involvement, 177, 180–82, 259–60, 323–24
issues
     behavioral, *136*, 148–52
     champions and, 141–46
     escalating, 383
     external, *136*, 159
     global market, *136*, 158
     leadership, 340–49, 378–84
     organizational and management, 153–57
     political, 136, 364–77, 422–24
     project and project management, 136–47,
          350–59, 378–84

special, 370–72
stakeholder management, 385–406

J

Jaffe, Dennis T., 345
Jick, Todd D., 234
joint ventures, *136*, 158

K

Kanter, Rosabeth Moss, 155
Kaplan, Robert E., 247
Keys, Barnard, 229
Kluckhohn, Florence Rockwood, 287
knowledge, *53*, 55–56, 175, 177–79, 370–72
Krackhardt, David, 247

L

leadership, 123, 340–49, 379–84, 391–92
Lee, C. A., 295
Lee, Raymond T., 244
Lewin, Kurt, 156
Liden, Robert C., 263
lobbyists. *See* consultants, external
Luthans, Fred, 285

M

MacGillivray, A., 263
Machiavellianism, *235*, 236–37

Madison, Dan L., 96, 242

management,
    change, *136*, 155–56
    clarification of goals, roles, and responsibilities,
        401–3
    conflict, 231–32, *275*, 280–81
    Dolphins and, *225*, 227–33
    Flounders and, *214*, 216
    good practices of, 290–94
    hierarchy of, 365–66
    issues, 153–57
    managing upward, 318–20, 364–77
    meeting, 279–80
    political behavior, 274–99
    senior, 296–97, 343–44
    Sharks and, *218*, 219–20
    stakeholder, 164–98, 314–26, 385–406, 418–21
    support of, 366–69
    See also champions; politics, management of;
        project management; stakeholders

Management by Projects. *See* MBP

Management Review, 156

Managing the Project Team, 22

Mannix, Elizabeth A., 254

Markus, M. L., 96

Martin, M. Dean, 285–86

Mayes, Bronston T., 96–97, 242

MBP (Management by Projects), 333–34

McKay, Harvey, 221, 223, 238

McKeen, Carol A., 255

McLaurin, James Reagan, 286

McShane, Steven L., 242

meetings, 112, 279–80

mentoring, *379*, 383–84, 394–95

mentoring program, *106*, 127–28, *332*, 338–39

mergers, *136*, 158

Miller, David, 167, 187

Mintzberg, Henry, 97, 234

Mitchell, Terence R., 263

momentum, *351*, 357–58

motivation *113*

Mudrack, Peter E., 237

Murray, Victor V., 234, 242

Myers, Dale, 214

N

name-dropping, 244

negative politics
    attributes of, 98–103, *99*
    behaviors that illustrate, 97–103, 242–68, *243,
        251, 258*
    emotional views as, 70–77, *71*, 417
    issues leading to,
            behavioral, *136*, 148–52
            external and global market, *136*, 158–59
            leadership, 340–49, 378–406
            organizational and management, *136*,
            153–57, 331–39
            political, 364–77
            project management, *136*, 136–40, 147,
            350–59, 364–406, *365, 379, 386*
    people engaged in, 234–37, *235*

preventing, *193*, 193, 195, *196*, 198
tolerance for, *136*, 156–57, *340*, 341
See also political behaviors; positive politics;
    Sharks
Nelson, Bob, 298
Nelson, Douglas, 167
networking, *225*, 228, *243*, 246–48
Newcombe, Robert, 185

O

objectives, *340*, 342
objectivity,
    analysis and, *106*, 126–27, *290*, 297–99, *332*,
        334–35
    reward system and, *258*, 264–65
obligation, *258*, 259–63
Offerman, L. R., 155, 241
Oliver, Mike, 167, 187
opponents, *394*, 395–96
organizational amnesia. *See* churning
organizational structure, 332–33
Organizing Projects for Success, 22
Owens, Stephen D., 286
ownership, *113*, 118, 262, 324, 345

P

panic, *71*, 71–73, *78*, 78, 80
Pearce, Jone L., 254

Pennings, J. M., 295

people
    engaged in negative politics, *235*
    with internal focus of control, *235*, 236
    with Machiavellian values, 235, 236–37
    power-hungry, 235
    unsatisfied and unhappy employees, *235*, 237
    withholding information, *235*, 236

Pfeffer, Jeffrey, 97

Pinto, Jeffrey K., 95, 97, 166, 211, 213, 217, 224, 227, 231, 239

PLC (project life cycle), 70–74, *71*, 82, 388
    phases of, 66–67, *68*

PLS (project life span), 66–69

PMBOK Guide, 27, 41, 46, 163–64, 167, 175, 177, *379*, 383, 385, 387, 409

PMI (Project Management Institute), 22, 26, 163, 167

PMO (Project Management Office), *78*, 81–84

policies, 290–94

political behaviors
    interpersonal level, 242–50
    managing, 274–99
    organizational, 258–68
    team environment, 250–57
    understanding, 242–68

political landscape, 210–37, 374–75, 418–21

Politically Naives. *See* Flounders

Politically Sensibles (PS). *See* Dolphins

political positions, 210–34

political structure, 173–75
politics
    basic concepts of, 47–61
    complexity of, 49–51
    components of, 51–3
    definition of, 47, 97
    dynamics of, 47–49, 97–131, 136–59, 164–98
    ideas about politics 47–48
    influence in a project environment, 68
    intensity of, *50*, 50–51
    level of according to PLS, 67–69
    managing, 314–26, 331–59, 364–407
    organizational, 136–59, 331–39
    PLS and, 66–69
    power and, 53–61
    project, 74–77
    project management and, 41–61, 66–84
    stakeholder management and, 164–98
    types of, 95–131
    See also negative politics; positive politics
Politics of Projects, The, 20–21
Porter, Lyman W., 96, 234, 242, 254
positive politics, 104–31, 224–25
    behaviors that illustrate, 243, 244–49, 251–55,
        258–64, 268
    proactive views as, 77–82
    Ten Commandments of, 105–31
    See also negative politics
power,
    acquiring, 57–59
    centralizing, 295–96
    champions and, 368–69
    corruption of, 295–96, 344–46

distributing, 345
effective use of, 380–81
formal, 53–59, 64, 99–100, 119, 147, 174–98
informal, 57, 58–59, 228, 381–82
ingredients of, 55–56
network, 51, 52
overall, 380–81
politics and, 53–61
seeking, 54–56
sources of, 55–57
struggles with, *136*, 150–51
total, 53–54
versus commitment to change, 187–90
versus ethics, 192–95
versus impact, *177*, 183–84
versus influence, *177*, 181–83
versus integrity, 195–98
versus interest, *177*, 179–81
versus knowledge, *177*, 178–79
versus predictability, 185–87
versus respect, 190–92

practices, good management, 290–95

praise and honor for nonparticipants, *71*, 75, *76*, *78*, *80*, 82

predictability, 185–87

prevention, 120

pride, 15, 118–19, 121, 262

proactivity, 77–84, *78, 80*, 417

problem-solving, 283–84, 403–5

project life cycle. *See* PLC (project life cycle)

project life span. *See* PLS

project management
  issues in, 136, 137–47, 350–59, 378–84
  politics and, 41–61, 66–84
Project Management Office. *See* PMO
project managers
  phases driven by, *78*, 79–81
  empowering, *340*, 344–46
  clarifying roles and responsibilities for, 353–54
projects
  analyzing and approving, 354–55
  Dolphins and, *225*, 227–33
  external, *136*, 139
  Flounders and, *214*, 216–17
  high-priority and high-profile, *340*, 346–47
  managing politics in, 364–406
  prioritization of, 136, 137–38, 290, 292–94
  quality of deliverables, 117–22
  risks of failure, 55–56
  selection of, **136**, 137–38, *290*, 291–92, 351–53
  Sharks and, *218*, 219–20
  stopping and restarting, 357–58
  See also benefits, project; PLC; PLS; project management; project managers; troubled projects
project sponsors, *78*, 79–81, 351, 353–54
punishment of the innocent, *71*, 74–75, *76, 78, 80,*
  81–82
Punzo, Richard A., 287

Q
quality, *106*, 117–22

R

Ragins, B. R., 255

reciprocity, *99*, 101

recognition, *290*, 297–99

Reframing Organizations, 21

regulations, 136, 159

relationships
Dolphins and, *225*, 228
strengthening working, *177*, 179, 184, *193*, 194

Renwick, Patricia A., 96, 242

resistance, *188*, 190, 245–46

resources
allocation of, *136*, 138–39, *290*, 294
scarce, *136*, 138, *290*, 294

respect, 190–92, *225*, 233

responsibilities, *351*, 353–54, *386*, 401–3

restructuring, *136*, 153–54

retribution, *99*, 102–3

revenge. *See* retribution

rewards, *57*, 58, 290, 297–98
nominating for, *258*, 268
systems, *106*, *113*, 126–27, *136*, 157, *332*, 334–35
See also favoritism

Rhinesmith, Stephen, 288

Roeder, Tres, 175, 177, 388–92

roles, 401–3
champions', 140–42, 346–47
project sponsors' and project managers', *351*,
353–54

Rosenfeld, Paul, 263

rules, *275*, 279–84, 315–16, *351*, 358–59

Russ, G. S., 234

S

Schneck, R. E., 295

Scott, Cynthia D., 345

search for the guilty, *71*, 74, *76, 78, 80*, 81

Seven Habits of Highly Effective People, 315

Shachtman, Tom, 264

Sharkey, Thomas W., 96

Sharks, 217–24, 374–75, 425–29

Singer, Michael, 319

skills, *53*, 55–56
    communication, *225*, 230–31
    interpersonal, 123, 141, 356–57
    influencing, 141, *146*
    leadership, 123, 348–49, 356–57
    people, *351*, 356–57, *379*, 383–84
    technical, 122, 356–57

Slocum, John W., 156

SMART goal-setting approach, 402

smoke screening, *258*, 265

solutions, win-win, *225*, 230, 405

staff shuffling, 154

stakeholders
    analyzing, 170–98, 388–89, 393–98
    attributes and characteristics of, 173–98, 389
    building alliances with, 390

categories of, 393–98
core, 169
external, 169
gaining support from, 390–91
identifying, 168–70, *386*, 387–88
internal, 169
managing
    issues in 385–406
    politics in, 164–98, 418–21
    strategies for, 388–92
    three truths of life, 314–26, 399–400
political behaviors of, 242–69
prioritizing, 175–98
unexpected, 169–70
Stebbins, M., 263
Steen, Sandra L., 242
Stern, Paul G., 264
Stevenson, William B., 254
Stewart, Edward C., 288
strategies, organizational, 106–13
structure, organizational, 332–33
subject matter experts. *See* consultants, external
Sundstrom, E., 255
support, *314*, 323–26, 400, 418
gaining,
    champions and, *365*, 366–69
    Dolphins and, *225*, 226, 228–29
    genuine involvement and, 323–26
    stakeholders and, *177*, 179, 184, *186*,187, 390–91
increasing, 188–89
lack of, *188*, 190, 197, 397

potential, 172, *190*, 191, 196

providing, *106*, 122–24, 277–78, 316–17

recognizing, *177*, 182, 197

See also training

survival, 74–75, *76*

swimming in a political pond, 212–13

Swim with the Sharks Without Being Eaten Alive,
221

synergy, 106–8, 116–17, 278

See also teamwork

T

tapping into the power lines, 165

Tasa, Kevin, 242

team norms, *275*, 276–78

teams, 250–57, 274–88, 316–18, 378–84

teamwork, *106*, 116–17, 220, 224–25, 316–18,
381–83

See also collaboration; synergy

Tedeschi, James T., 263

training, *106*, 122–25, *351*, 356–57

See also support

Training Management Corporation, 287

Trompenaars, Alfons, 286

troubled project

dynamics of, 70–84, 415–17

emotional views of, 70–77, 71, 417

phases driven by PLC and project dynamics,
70–74

phases driven by PMO, *78*, 81–84

phases driven by project manager and sponsor, *78*, 78–81
phases driven by project politics, *71*, 72–77
proactive views of, 77–84, *78, 80,* 417
trust, *190*, 191–92, *214*, 216, 393–98
Turner, Charles H., 286
Tzu, Sun, 232

U

Ury, William L., 230

V

values
    Machiavellian, *235*, 236–37
    moral, *290*, 296–97
    organizational, *275*, 276–78, *340*, 343–44
vision, *340*, 342

W

Walker, Derek H. T., 165
Watt, Adrienne, 171
Wellins, Richard S., 345
What's in It for Me. *See* WIIFM
What's in It for Them. *See* WIIFT
WIIFM (What's in It for Me), 229, *314*, 321, 373, 399
WIIFT (What's in It for Them), 184, 229, 322–23, 386, 400
    champions and, 142–144, *146, 365*, 367
    managing upward and 373, 377

Wikipedia, 213, 217, 224
Wilson, Jeanne M., 345
Woodman, Richard W., 156

The Art of Positive Politics

About the Author

Vijay K. Verma is an internationally renowned speaker and author. He wrote a three-volume series on the Human Aspects of Project Management published by the Project Management Institute (PMI): *Organizing Projects for Success, Human Resource Skills for the Project Manager,* and *Managing the Project Team.* Mr. Verma received the 2009 PMI Fellow Award (one of the highest and most prestigious awards presented by PMI), the 1999 PMI David I. Cleland Project Management Literature Award (for his book *Managing the Project Team*), and the 1999 PMI Distinguished Contribution Award for sustained and significant contributions to the project management profession.

Mr. Verma has given many keynote presentations at various conferences. He has authored and presented many papers at national and international conferences on the Human Aspects of Project Management and Managing Cross-Cultural Teams. He has presented several workshops on project management in the United States, Canada, Europe, Australia, South Africa, China, and India with participants from a wide range of industries. More than four thousand professionals working in project management have attended his presentations to enhance their skills and confidence in managing projects effectively.

Mr. Verma worked for more than thirty-nine years in project management at TRIUMF (TRI University Meson Facility), Canada National Research Laboratory. Here, he provided project management services for projects varying in size, complexity, and diversity. Mr. Verma served as president of the West Coast BC Chapter (1988–1989). He is a professional engineer, and he holds an MSc in Electrical Engineering from DalTech, Halifax, and an MBA from the University of British Columbia, Vancouver, Canada. He lives in Vancouver with his wife.

Do you want more?

If you enjoyed this book and want more
information on this topic, the author is available
for training and consulting engagements
through Procept Associates Ltd.

Procept delivers management training and
consulting services across North America and
around the world. Its award-winning trainers
and consultants deliver services customized to
your industry or your own company's challenges,
processes, and culture.

Find out more at:

www.Procept.com

+1 (416) 693-5559 (Direct)

1-800-261-6861
(Toll-free in Canada & USA only)

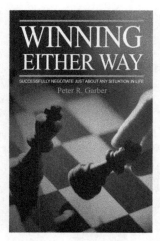

Winning Either Way: Successfully Negotiate Just About Any Situation in Life

by Peter R. Garber

There are no sure bets in life. To succeed in life, we need a philosophy - a mindset - a way of looking at things from a different perspective. *Winning Either Way* teaches you how to look at the world through a different lens and to move away from the win/lose perspective that is so common, finding ways to get something that is of value to you in just about any situation.

Life is full of unexpected opportunities but you have to be able to see them. If life gives you lemons, you should go into the lemonade business. Losing in the traditional sense isn't always the worst thing that could happen. Change can bring unexpected opportunities. This book will teach you how to prepare yourself to spot these opportunities so that you can take advantage of them.

This book teaches negotiating strategies and techniques that can prepare you to always come out with what you need in any negotiating situation. No matter how the negotiations go, you'll "win either way."

Project: Impossible — How the Great Leaders of History Identified, Solved, and Accomplished the Seemingly Impossible — and How You Can Too!

by Michael Dobson

Nothing is impossible if you have unlimited time, resources, and flexible objectives. Project managers never find themselves in such a situation. Our projects are impossible if they can't be done within the constraints...but sometimes there's a way around even the most challenging barrier.

What can you do when the situation looks hopeless? In this exciting journey through history, you'll learn how the greatest leaders and project managers of the past took on impossible challenges...and succeeded.

- When he wanted to be first to fly nonstop to Paris, Charles Lindbergh was up against competitors with more funding, more experience, and over a year's head start. His strategy? Rethink the thresholds of risk.

- For mission director Gene Kranz, the odds against a successful rescue of Apollo 13 were daunting at best. His strategy? The Kranz Dictum, a powerful strategy to deal with crises even before they occur.

In PROJECT: IMPOSSIBLE, you'll learn a step-by-step methodology to succeed when facing even the most difficult projects. You'll learn Dobson's Laws of Project Management and discover the Godzilla Principle. From redefining the problem to challenging the project parameters, you'll know how to attack a seemingly impossible project...and get the job done.

Top-Gun Project Managers: 8 Strategies for Reaching the Top of the PM Profession
by Richard Morreale

Ever wonder why some people's careers rocket upwards while yours feels stuck, with you repeating the same drudgery day after day? If you are a project manager (or are interested in becoming one) then this book was written just for you. It shares eight strategies that you can use when plotting your career path -- the trajectory of your own rocket -- to help you reach the stratospheric levels of the profession.

These eight strategies are not just high level theory. The author of this book, Richard Morreale, has put them into practice successfully in his own career. Morreale is one of the top project managers in the world, specializing in turning around some of the nastiest, largest, troubled projects you will ever find; in fact, Morreale is often called "the Red Adair of project management" after the famous engineer who specialized in putting out oil well fires. Morreale's career has spanned a wide range of projects from working as part of the Apollo Program Team, helping to put men on the moon (and getting them back), to working as part of the team that computerized the UK Income Tax System. He also led the rescue and delivery of a $450M program for the 43 Police Forces in England and Wales and directed programs of work for some of the largest companies in the world.

Read this book to learn how you too can copy Morreale's career success -- with these eight strategies, the sky's the limit!

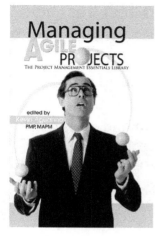

Managing Agile Projects

Edited by Kevin Aguanno

Are you being asked to manage a project with unclear requirements? High levels of change? A team using Extreme Programming or other Agile Methods?

If you are a project manager or team leader who is interested in learning the secrets of successfully controlling and delivering agile projects, then this book was written for you. From learning how agile projects are different from traditional projects, to detailed guidance on a number of agile management techniques and how to introduce them onto your own projects, this book contains the insider secrets from some of the industry experts - the visionaries who developed the agile methodologies in the first place.

Chapters focus on topics critical to the success of projects facing changing requirements and seemingly impossible deadlines. Chapters cover topics such as engineering unstable requirements, active stakeholder participation, conducting agile meetings, extreme testing, agile documentation, and how to use agile methods under fixed price contracts. The book also provides information to help you plan your agile projects better to avoid some common pitfalls introduced by the fast pace and concurrent activities common to agile development methods.

This book will show you the tricks to keeping agile projects under control.

Today is a Good Day: Attitudes for Achieving Project Success
by Alfonso Bucero

There is an old saying that "your attitude determines your altitude." People with a positive, constructive attitude achieve much greater success than those with neutral or negative attitudes.

This book explores the impacts of attitude on project success. Examining the importance of a project manager's attitude in motivating a team, dealing with project stakeholders, and overcoming issues, the book contrasts the outcomes of both positive and negative attitudes.

Full of case studies illustrating the key concepts, the book shows how project managers can change their way of thinking, plan for success, take advantage of the power of commitment, convert project issues into opportunities, communicate more effectively, deal effectively with negativity and fear, and effectively use persistence to increase the chances for project success.

With the right positive attitude, problems seem to melt away. Don't get caught up in negativity surrounding project issues and risks, because... Today is a Good Day!

The Keys to Our Success: Lessons Learned from 25 of our Best Project Managers

Compiled by David Barrett & Derek Vigar

Project managers are keen to learn from the best. So we asked the top experts in the field a straightforward question: "What is your best piece of advice for success in project management?

The result is this book—a collection of their best stories, lessons, and takeaways. 25 different industry leaders make compelling cases for why their "key" will help influence your project success: the seven bullets of highly effective project managers; why leadership must be taken, not given; the importance of becoming project "business-savvy"; ways to generate meaningful client ownership; how great project managers make it fun; and 20 other differentiators that have helped these industry leaders stand out.

If you are interested in differentiating yourself and boosting your career, then this book is a fantastic opportunity to connect with trusted mentors, read their honest advice, and leverage these keys to success in your own practice.

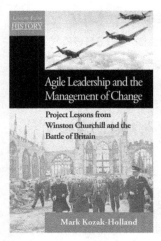

Agile Leadership and the Management of Change: Project Lessons from Winston Churchill and the Battle of Britain

by Mark Kozak-Holland

There was a poll conducted in Britain that asked who people thought was the most influential person in its history. The winner: Winston Churchill.

What set Churchill above the others was his leadership qualities: his ability to create and share a powerful vision, his ability to motivate the population in the face of tremendous fear, and his ability to get others to rally behind him and quickly turn his visions into reality. By any measure, Winston Churchill was a powerful leader. What many don't know, however, was how Churchill acted as the head project manager of a massive change project that affected the daily lives of millions of people, to restructure the British military, government, and even the British manufacturing sector to support his efforts to rearm the country and get ready for an imminent enemy invasion in early 1940

Learn about Churchill's change management and agile management techniques and how they can be applied to today's projects.

The Elusive Lean Enterprise: Why Companies Struggle to get the Full Benefits from Lean

by Keith Gilpatrick and Brian Furlong

In today's fast-paced and volatile business environment, customers are demanding increased flexibility and lower cost, and companies must operate in a waste-free environment to maintain a competitive edge and grow margins. Lean Enterprise is the process that companies are adopting to provide superior customer service and to improve bottom line performance.

This book is designed to help guide you through the Lean transformation and avoid common pitfalls. Learn from the mistakes of others and avoid the trials that often kill the initiative. Find out why you must change, how to change, and how to institutionalize the process. Understand the costs of outsourcing or going offshore and compare these to the Lean alternative.

For those companies that fail to commit to the process and truly change the culture, a Lean Enterprise will remain elusive.

CPSIA information can be obtained
at www.ICGtesting.com
Printed in the USA
LVHW040156140919
631067LV00002B/2